Civil Engineering Specification

By the same author

Civil Engineering Quantities
Building Quantities Explained
Municipal Engineering Practice
Building Economics
Building Technology
Building Maintenance
Outdoor Recreation and the Urban Environment
Planned Expansion of Country Towns (published by George Godwin)

Civil
Engineering Specification

Ivor H. Seeley

B.Sc., M.A., Ph.D., F.I.Mun.E., F.R.I.C.S., F.I.Q.S., M.I.O.B.
*Head of Department of Surveying
and Dean of the School of Environmental Studies, Trent Polytechnic,
Nottingham*

Second Edition

First published 1968
Second edition 1976

Published by
THE MACMILLAN PRESS LTD
London and Basingstoke
Associated companies in New York Dublin
Melbourne Johannesburg and Madras

ISBN 0 333 04227 1 (hard cover)
0 333 19870 0 (printed-case)

Printed in Great Britain by
THE SCOLAR PRESS LTD

Contents

Preface   vii

Acknowledgements   viii

1 CIVIL ENGINEERING CONTRACTS   1

Contract Documents: Form of Contract; General Conditions of Contract;
Specification; Bill of Quantities; Contract Drawings; Form of Tender –
I.C.E. Conditions – I.Struct.E. Conditions – Types of Contract.

2 PURPOSE AND ARRANGEMENT OF AND SOURCES OF   16
   INFORMATION FOR SPECIFICATIONS

Functions of Specification – General Arrangement – Drafting of
Specification Clauses – Sources of Information – British Standards –
Codes of Practice.

3 GENERAL CLAUSES IN SPECIFICATIONS   26

General Contractual Matters – Legal Provisions – General Matters
affecting Cost of Job – Use of Site – Temporary Works – Materials and
Workmanship Requirements – General Working Requirements – Records.

4 SPECIFICATION OF EARTHWORK   49

Site Investigation and Clearance – Excavation, Fill and Disposal –
Keeping Excavations free from Water – Timbering – Tunnel Work –
Cofferdams – Dredging.

5 SPECIFICATION OF CONCRETE WORK   65

Materials – Concrete Work, including mixing, placing, joints, curing
and testing – Reinforcement – Shuttering.

6 SPECIFICATION OF BRICKWORK, MASONRY AND   88
   WATERPROOFING

Brickwork: Bricks; Mortars; Brickwork; Damp-proof Courses; Faced
Brickwork, etc.
Masonry: Dressed Stonework; Rubble Walling; Special Stonework;
Fixings; Cast Stonework.
Waterproofing: Asphalt; Bitumen Sheeting.

7 SPECIFICATION OF PILING   103

Concrete Piles, including materials, shoes, casting, curing, trial piles,
handling, pitching, driving and lengthening. Timber Piles, including
timber, creosoting or tarring, shoes and rings, pitching, driving and
cutting off heads. Steel Sheet Piling, including driving, cutting and
drilling piles.

8   SPECIFICATION OF IRON AND STEELWORK   114

Structural Steel – Fabrication – Erection – Bolting – Riveting –
Welding – Measurement – Testing – Ladders – Guardrails – Open
Steel Flooring – Painting – Wall and Roof Sheet Coverings: Asbestos
Cement; Aluminium; Corrugated Steel.

9   SPECIFICATION OF TIMBERWORK   126

Timberwork in Wharves and Jetties: Quality of Timber; Workmanship;
Fender Piles, Rubbing Pieces, Walings, Braces, Guardrails and Decking;
Tarring and Creosoting; Bolts and Nuts; Measurement; Equipment –
Steps, Footbridge and Scumboards – Joinery: Quality of Timber;
Windows; Doors; Miscellaneous Joinery Work; Painting.

10   SPECIFICATION OF ROADS AND PAVINGS   142

Materials – Road Bases, including lean concrete and soil cement –
Flexible Road Construction: Tarmacadam; Bitumen Macadam; Rolled
Asphalt; Cold Asphalt; Surface Dressing – Concrete Carriageway
Construction – Kerbs, Channels, etc. – Road Gullies – Electric Cable
Ducts – Footpaths: Tarmacadam; Asphalt; Flagged; 'Insitu' Concrete –
Grass Verges – Chain-link Fencing.

11   SPECIFICATION OF SEWERS AND DRAINS   171

Materials – Excavation – Pipelaying, including concrete protection and
testing of pipes – Manholes: Brick and Precast Concrete; Associated
Metalwork – Tunnel and Shaft linings: Cast Iron and Precast Concrete
Segments; Jointing; Grouting; Concrete Lining – Ventilating Columns –
Sewage Works Filters, Scumboards and Screens.

12   SPECIFICATION OF PIPELINES   200

Materials, including valves – Pipelaying under varying conditions –
Watercourse and River Crossings – Reinstatement of Trench Surfaces –
Testing and Sterilisation of Pipes – Valve Chambers: Sluice Valves;
Air Valves; Washouts; Hydrants.

13   SPECIFICATION OF RAILWAY TRACKWORK   215

Preliminary Work – Track Materials: Ballast; Sleepers; Rails; Fishplates;
Rail Fixings; Separators; Bearing Plates; Chairs and Keys; Laying
Permanent Way: Laying Ballast; Laying Track; Points and Crossings;
Measurement of Railway Work.

Appendix I   List of British Standard Codes of   229
Practice relating to Civil Engineering Work

Appendix II   List of British Standards relating to   231
Civil Engineering Work

Appendix III   Typical Programme of Works covering   237
Tunnels and Shafts to a Circulating Water System to a
Power Station

Appendix IV   Metric Conversion Table   239

Index   241

Preface

THIS book is primarily concerned with the drafting of specification clauses for civil and municipal engineering work. The specification on a civil engineering contract constitutes a contract document and its provision is essential if the material and workmanship requirements of the job are to be adequately detailed.

The book sets out to indicate how a comprehensive and yet straightforward specification can be produced. The specification acts as one of the principal lines of communication between the Engineer and the Contractor and thus requires the exercise of considerable care and skill in its preparation.

The contents of this book should be helpful to engineers when compiling specifications and also of value to students who are studying the subject for degrees, diplomas and professional examinations in civil and municipal engineering.

Units of measurement, weight and pressure have been converted to their metric equivalents and the imperial units are shown in brackets. Appendix IV contains a metric conversion table which readers may find useful when preparing their own specifications on the metric system.

Nottingham, 1976   I. H. SEELEY

ACKNOWLEDGEMENTS

THE author expresses his indebtedness to the various consulting and municipal engineers from whom over the years he has received valuable experience and guidance in the drafting of specifications.

Grateful thanks are also due to P. J. Edmonds and T. E. Blackall of the publishers, for abundant help and consideration during the production of the first edition of the book, and to Malcolm Stewart, for the new edition.

CHAPTER ONE

Civil Engineering Contracts

THIS Book is mainly concerned with the drafting of specifications for civil and municipal engineering work, but it is felt desirable to begin by considering the relationship of the specification to other contract documents, particularly the Conditions of Contract.

With any major civil engineering project, it is necessary for the engineer to prepare a set of comprehensive contract documents. These are all binding upon the contractor, who must pay full regard to their contents when tendering for a job and throughout the contract period, when the work is under way.

CONTRACT DOCUMENTS

There are usually six contract documents for the larger civil engineering jobs, as listed below, although with very small contracts it is conceivable that a bill of quantities may be omitted.

(1) Form of Contract.
(2) General Conditions of Contract.
(3) Specification.
(4) Bill of Quantities.
(5) Contract Drawings.
(6) Form of Tender.

The nature and uses of each of these documents are now described.

(1) Form of Contract

The Form of Contract constitutes the formal agreement between the promoter and the contractor for the execution of the work in accordance with the other contract documents. It is usually covered by the Form of Agreement incorporated in the General Conditions of Contract for Works of Civil Engineering Construction. (See reference 1 at the end of the chapter.)

In the Form of Agreement, the contractor covenants to construct, complete and maintain the works in accordance with the contract, and the promoter or employer covenants to pay the contractor at the times and in the manner prescribed by the contract. These are the basic requirements of any civil engineering contract.

(2) General Conditions of Contract

The General Conditions of Contract define the terms under which the work is to be undertaken, the relationship between the promoter or employer, the engineer and the contractor, the powers of the engineer and the terms of payment. For many years it had been considered desirable to establish a standard set of generally recognised conditions which could be applied to the majority of civil engineering contracts.

In 1945, the Institution of Civil Engineers, in conjunction with the Federation of Civil Engineering Contractors, issued a standard set of General Conditions of Contract for use in connection with Works of Civil Engineering Construction (1). The Association of Consulting Engineers was concerned with the preparation of later editions of this document, in addition to the two bodies previously mentioned. Furthermore, other sets of conditions have been specially prepared to cover civil engineering works which are to be performed overseas (2).

The Institution of Structural Engineers has issued a set of standard conditions for use on structural engineering contracts (3) and the main clauses will be considered later in this chapter.

For building work, it is customary to make use of the standard conditions issued under the sanction of the Royal Institute of British Architects and various other bodies, and generally referred to as the J.C.T. Conditions (4). There are alternative forms for use where quantities do or do not form part of the contract and there are, in addition, sets of conditions specially devised for use on local authority contracts. Practice notes are issued from time to time to clarify doubtful points.

Where a contract is very limited in extent and the use of the standard comprehensive set of conditions is not really justified, an abbreviated set of conditions, often worked up from the appropriate set of standard conditions, is frequently used.

With certain specialised classes of civil engineering work, the responsible authorities have seen fit to introduce a number of clauses which modify or supplement the standard clauses of the I.C.E. Conditions. Typical examples are the clauses prepared by the Central Electricity Generating Board for use on power station contracts, and the clauses introduced by the Ministry of Transport in connection with contracts for roads and bridges.

(3) Specification

The specification amplifies the information given in the contract drawings and the bill of quantities. It describes in detail the work to be executed under the contract and the nature and quality of the materials and workmanship. Details of any special responsibilities to be borne by the contractor, apart from those listed in the general conditions of contract, are often incorporated in this document. It may also contain clauses specifying the order in which various sections of the work are to be carried out, the methods to be adopted in the execution of the work, and details of any special facilities that are to be afforded to other contractors.

In *Civil Engineering Procedure* (5), issued by the Institution of Civil Engineers, it is recommended that the specification should require tenderers to submit an outline programme and a description of proposed methods and temporary works with their tenders.

The specification will always be a contract document on civil engineering contracts, while in the case of building contracts, operating under the J.C.T. form of contract, it will be only a contract document if there is no bill of quantities or when it is specifically made a contract document under the particular contract.

(4) Bill of Quantities

A bill of quantities consists of a schedule of the items of work to be carried out under the contract, with quantities entered against each item. Quantities of civil engineering work are normally measured in accordance with the Standard Method of Measurement of Civil Engineering Quantities (6), whereas most building work is measured in conformity with the Standard Method of Measurement of Building Works (7). For road and bridge works, the code issued by the Department of the Environment (11) may prove useful, although it conflicts with some of the provisions of the Standard Method of Measurement of Civil Engineering Quantities (now CE SMM) (13).

One of the primary functions of a civil engineering bill of quantities

3

is to provide a basis on which tenders can be obtained and, when the bills are priced, they afford a means of comparing the various tenders received, both as regards totals and individual rates. After the contract has been signed, the rates in the priced bill of quantities can be used to assess the value of the work as executed, and help in the preparation of interim statements and calculation of bonus.

The quantities should be as accurate as possible, although on civil engineering contracts the work is invariably remeasured on site, whereas on building jobs the remeasurement of work on site is mainly limited to substructural and drainage work and work which is the subject of variations. Civil engineering billed descriptions are kept brief, with frequent references to specification clauses, whereas in building work detailed descriptions appear in the bill. In all cases, billed descriptions must indicate clearly the nature and scope of the work covered.

The unit rates entered by contractors against items in bills of quantities normally include for overheads and profit, and in civil engineering bills the rates also have to include for the majority of items of temporary work, such as trench timbering, keeping excavations free from water, and levelling and ramming bottoms of excavations. Special items of temporary work, such as the construction of a cofferdam around a sewer outfall into a river, may call for special billed items.

The actual arrangement of a civil engineering bill of quantities varies with the type of work being measured. For instance, a bill for a sewage disposal works will usually be split into sections, each covering a component part of the works, such as sedimentation tanks and biological filters. Within each section, the billed items will normally follow the order of subsections contained in part IV of the Standard Method of Measurement of Civil Engineering Quantities. In building work, the bill is subdivided into trades or work sections.

Where the same constructional work is to be performed under different conditions, separate billed items should be provided to give the contractor the opportunity to enter different prices against them. Typical examples are a reinforced concrete slab in the base of a pump well below ground and a similar slab in a water tower tank 21 m (70 ft) above ground. Where the quantity of work is uncertain, such as making up soft spots in a road formation, then these should be listed as 'provisional'.

Where 'prime cost' sums are entered in the bill to cover the cost of materials to be supplied or work to be done by sub-contractors, the main contractor must be given the opportunity of adding sums to cover profit and fixing of materials or attendance on sub-contractors. General provisional sums are frequently included in bills of quantities to cover

4

contingencies and additional works that may arise during the course of the contract, due to site conditions or changes in design.

The Standard Method of Measurement of Civil Engineering Quantities permits the use of a system of comprehensive measurement for repetition work, mainly composite work of a uniform type of construction. A single billed item can be used, although several classes of workmanship and materials may be involved. Typical examples of work where this method can be employed are retaining walls, tunnel work and pipelines. In each case a subsidiary bill should be included in the description column listing, in the recognised units of measurement, the quantities of the component materials and work required to provide a unit of length, e.g. one metre (linear yard) of tunnel lining. Where the 'Variation of Price (Labour and Materials) Clause' operates, the contractor is permitted to claim the increased costs which occur after the date of tender, for labour, materials and consumable stores, which are used on the contract. It is the usual practice in these circumstances to include a schedule of basic rates at the end of the bill of quantities, in which the contractor can enter the basic prices on which his tender is based. In some cases the principal materials are entered in the schedule when the bill is being prepared, but the contractor can enter any other materials if he so wishes.

To secure uniformity of measurement, recognised units of measurement and general rules and principles appertaining to the work of measurement are detailed in the Standard Method. The application of these rules to the measurement of the different classes of civil engineering work is illustrated in a book by the author (9), which also describes the processes involved in the preparation of a bill of quantities.

(5) Contract Drawings

Contract drawings depict the nature and scope of the work to be carried out under the contract. They must be prepared to a suitable scale and be in sufficient detail to permit a contractor to price the bill of quantities and to carry out the work satisfactorily. For instance, sewer plans are frequently drawn to a scale of 1 : 2000 (1/2500) or 1 : 500 (1/500), with sections prepared to an exaggerated scale, such as 1:2000 (1/2500) horizontal and 1:500 ($\frac{1}{40}$ in. to 1 ft) vertical. Road plans are sometimes drawn to 1:500 (1/500) scale. Scales of 32 or 64 ft to 1 in. have been used to an increasing extent (multiples of $\frac{1}{8}$ in. scale) but the 1:500 scale will now probably become universal. Manhole and gully details are usually drawn to a scale of 1:20 ($\frac{1}{2}$ in.:1 ft).

All available information as to the topography of the site, the nature of the ground and the water-table level, should be made accessible to contractors tendering for a job. Existing and proposed work must be clearly distinguished on the drawings. For instance, old and new sewers and other services can be depicted in different colours or different types of line. With alterations to buildings it is often preferable to prepare separate plans of old and new work.

All drawings should contain an abundance of descriptive and explanatory notes which should be clearly legible and free from abbreviations. Ample figured dimensions should be inserted on the drawings to ensure maximum accuracy in taking-off quantities and in setting-out the constructional work on the site.

Materials shown in section are best coloured or hatched for ease of identification, using the notation in the relevant British Standard (10). Guidance is also given in the British Standards on the use of lines for various purposes: for instance, dimension lines are to be thin and continuous with the dimension placed above the line and along it, and readable from the bottom or right-hand edge of the drawing.

There are nine recommended engineering drawing sheet sizes, ranging from 2 m × 1 m (72 in. × 40 in.) to 250 mm × 200 mm (10 in. × 8 in.) (sheet sizes as cut from roll) in B.S. 3429. Where a civil engineering department keeps a stock of ready-printed negative sheets, it will probably be decided to limit the number of sheet sizes to about 4–6. Recommendations are also included as to the layout of drawing sheets and numbering of drawings (10). It is good practice to keep a card index of drawings and to enter on drawings the date and nature of any amendments.

One set of contract drawings will be used by the contractor on the site, and these will probably be supplemented by further detailed drawings during the course of the contract. Prints are obtainable on paper or linen, the latter being more suitable for rough usage on the site. Dyeline prints are not true to scale and can give rise to errors in scaling dimensions.

Information on drawings over-rides that given in the specification which, in its turn, over-rides the bill of quantities, and all three are over-ridden by the Conditions of Contract.

(6) Form of Tender

The Form of Tender constitutes a formal offer by the contractor to execute the contract works in accordance with the contract documents for the contract price or tender sum. It usually incorporates the contract period within which the contractor is to complete the works.

The form of tender now generally used for civil engineering contracts is the form incorporated in the General Conditions of Contract for use in connection with Works of Civil Engineering Construction (1). This form of tender provides for a 'bond' amounting to 10 per cent of the tender sum. The contractor is generally required to enter into a bond, whereby he provides two sureties or a bank or insurance company who are prepared to pay up to 10 per cent of the contract sum if the contract is not performed satisfactorily.

The appendix to this form of tender covers the amount of the bond, minimum amount of third party insurance, time for completion, amount of liquidated damages, period of maintenance, percentage for adjustment of prime cost sums, percentage of retention, limit of retention money, minimum amount of interim certificates and the time within which payment is to be made after a certificate is issued.

CONDITIONS OF CONTRACT

The most important provisions of the two main standard sets of conditions used on civil engineering contracts are now summarised.

I.C.E. Conditions

The duties of the engineer's representative are stated in clause 2: 'to watch and supervise the construction, completion and maintenance of the Works'. Furthermore, the engineer may delegate in writing any of his powers to his authorised representative.

The contractor is entitled to two sets of drawings (clause 6) and he is required to examine the site before submitting his tender (clause 11). Clause 14 requires the contractor to submit a detailed programme, showing the order and method of carrying out the work. The contractor shall not assign any part of the contract or sub-let any part of the works without the written consent of the employer or engineer respectively (clauses 3 and 4).

The contractor is responsible for the true and proper setting-out of the works and for the provision of all necessary instruments and labour (clause 17). Under clause 16, the engineer may require the contractor to remove from the works any person who misconducts himself or who is incompetent or negligent.

The contractor is also responsible for the care of all permanent and temporary works (clause 20), insurance of the works (clause 21),

7

damage to persons and property (clause 22), giving of notices and pay-ment of fees, compliance with relevant statutes and regulations (clause 26), providing reasonable facilities for other contractors on the site (clause 31) and clearance of the site on completion (clause 33). It will be noted from clause 30 that the passage of traffic to and from the site is to be organised, as far as possible, so as to reduce to a minimum all claims for damage caused by extraordinary traffic.

Clause 34 provides for the payment of wages and observance of hours and conditions of labour not less favourable than those established for the trade or industry in the district where the work is carried out.

The requirements with regard to costs of samples and tests (clause 36) and the inspection and removal of improper work and materials (clauses 38 and 39) are very important. No work shall be covered up without being approved by the engineer, and the engineer has power to order, in writing, the removal of improper materials or work.

Clause 40 gives the engineer the right to suspend the progress of the works but, in the majority of cases, any extra cost incurred by the contractor as a result of the suspension will have to be borne by the employer.

The contractor is required to begin the works within 14 days after receipt of the engineer's written order to proceed (clause 41). Whilst clauses 43, 44, 47 and 48, relating to time for completion, liquidated damages for delay and certificate of completion, all have far-reaching effects.

The contractor is required to make good any defects arising from the use of materials or workmanship not in accordance with the contract appearing during the maintenance period (clause 49).

The engineer may alter the extent or character of the works, by orders in writing, without invalidating the contract, and the additional or amended works shall be valued at billed rates, as far as practicable (clause 51). The engineer is given the power to fix rates for varied work, and provision is also made for carrying out certain work on a daywork basis (clause 52). 'Daywork' is the method of valuing work on the basis of the time spent by the workmen, the materials used and the plant employed, plus a percentage to cover overheads and profit. This method is used when it is impracticable to value work at the billed rates and the only satisfactory method of evaluation of the work is on a daywork basis. The Standard Method of Measurement of Civil Engineering Quantities (6) details three ways of valuing work on this basis.

Clause 53 vests the ownership of all constructional plant, temporary work and materials on the site in the employer, and these cannot be removed by the contractor without the written consent of the engineer.

The quantities are to be measured in accordance with the Standard Method of Measurement of Civil Engineering Quantities, and the work as executed is to be measured on completion (clauses 56 and 57).

Under clause 58, both selected suppliers of goods and selected sub-contractors on a civil engineering contract are referred to as 'nominated sub-contractors'.[1] For prime cost items the engineer is empowered to order the contractor to employ a sub-contractor nominated by the engineer for the execution of any work or supply of any goods or services involved. The contractor when required by the engineer shall produce all quotations, invoices and the like relating to work carried out by nominated sub-contractors.

The contractor submits to the engineer a monthly statement of the estimated value of permanent work executed and, if it exceeds the minimum amount for interim certificates, the engineer issues a certificate authorising payment to the contractor, less retention money. The contractor is entitled to interest on overdue payments based on the minimum lending rate plus $\frac{3}{4}$ per cent (clause 60).

If the contractor becomes bankrupt or fails to perform his obligations under the contract, he becomes liable to expulsion from the site (clause 63). The engineer is authorised to settle any disputes arising under the contract, but if the contractor is dissatisfied with his decision, he can take the matter in dispute to arbitration (clause 66).

The contract price fluctuation clause can be incorporated where it is not to be a 'fixed price' contract, when the extra costs will be computed from labour, plant and materials cost indices.

Institution of Structural Engineers, Conditions

The same provisions as in the I.C.E. Conditions apply as to assignment and sub-letting (clause 2). All work, including approximate items, shall be contained in the bill of quantities, otherwise the contractor can claim additional payment (clause 3). Under clause 4, the contractor is entitled to two copies of the drawings and other contract documents.

The contractor is to be given full possession of the site within a stated number of days of the signing of the contract and he is to complete the works by a specified date. This latter date may be extended by virtue of one or more of the five causes mentioned in this clause. If the contractor fails to complete the works within the specified period, suitably extended where appropriate, he shall become liable for payment

---

[1] 'Nominated sub-contractors' on civil engineering contracts are specialist tradesmen, merchants and others carrying out work or supplying goods, for which provisional or prime cost sums are included in the bill of quantities. These sub-contractors are selected by the engineer on behalf of the employer, but they receive their instructions and payment from the main contractor. In building contracts, the person nominated for the supply of goods is described as a 'nominated supplier'.

9

of a specified weekly sum as liquidated and ascertained damages (clause 6).

Clause 9 contains similar testing provisions to those contained in the I.C.E. Conditions, and similar powers with regard to the removal from the site of unsatisfactory persons (clause 10). Under clause 11, the contractor is responsible for paying all fees and giving all notices legally demandable, except that he can recover the cost of fees from the employer if they are not covered by a provisional sum in the bill of quantities. The contractor is to supply the engineer with a detailed programme showing the methods, plant and order to be adopted (clause 12).

Under clause 13, the contractor is obliged to pay rates of wages and observe hours of labour not less favourable than those commonly recognised by employers and trade unions in the district where the work is being performed. Furthermore, the billed rates are to include for welfare requirements and all labour costs.

The contractor is entitled to recover the amount of increases in the cost of wages and materials occurring during the period of the contract. Similarly, the employer will be entitled to the benefit of any decreases that occur (clause 14).

By virtue of clause 15, the contractor is responsible for the care of the works from their commencement to the end of the maintenance period and shall, at his own expense, make good any damage or loss unless this is caused by certain specified acts. The contractor is made liable for all injuries or damage caused to any person or property due to the construction of the works, with certain specified exceptions (clause 16). The contractor is also to insure all works against loss or damage by fire and other specified risks (clause 17).

Materials, plant and temporary works on the site vest in the employer (clauses 18 and 19).

The contractor is obliged to provide adequate supervisory staff on the job (clause 20). The resident engineer is empowered to order the removal of condemned work or materials and to order variations (clause 21). It is not specifically mentioned that these orders are to be in writing. Under clause 22, billed rates are to include all liabilities and obligations listed in the General Condition of Contract and Specification. Varied work is to be valued at billed rates as far as practicable (clause 23). Clause 24 introduces the quantity surveyor for measurement and valuation of work, whereas he was not mentioned in the I.C.E. Conditions. Clause 25 requires bills of quantities for structural work to be prepared in accordance with the Standard Method of Measurement of Building Works, but in practice the civil engineering code of measurement is often used.

The contractor has the same responsibilities for setting-out the works as under the I.C.E. Conditions (clause 27). The contractor is to afford reasonable facilities on the site to contractors and workmen engaged by the employer (clause 28) and allow the employer and engineer access to the site and appropriate workshops (clause 29). The contractor is to give notice to the engineer before any work is covered up (clause 30).

Under clause 32, the contractor is entitled to extra payment for excavation in rock, running sand or artificial obstructions. The contractor shall not execute permanent work at night or weekends, without the written consent of the engineer, except in special circumstances (clause 33). Where the contractor is required to suspend the works, he will normally be recompensed for the additional cost involved (clause 34).

The contractor is entitled to $2\frac{1}{2}$ per cent cash discount on sub-contractors' work and 5 per cent on nominated suppliers' work. If a contractor fails to pay nominated sub-contractors or suppliers the cost of work or materials included in previous certificates, the employer may pay the accounts direct and deduct the amounts paid from the sums due to the contractor (clause 36). This clause follows the procedure laid down in the J.C.T. Conditions.

The engineer is given the same power to vary the works as in the I.C.E. Conditions. Varied work shall be valued at billed rates wherever possible, and on a daywork basis if the extra work cannot be satisfactorily measured and valued (clause 37).

Under clause 38, the contractor is responsible for furnishing the engineer monthly with details of the amount of work performed and materials on site. The engineer is to issue his certificate within 14 days of the receipt of this statement. One-half of the retention money is to be paid to the contractor on certified completion and the remainder at the expiration of the maintenance period, provided that all defects have been remedied. This clause further provides that if the employer fails to honour certificates within 14 days of issue or to settle the final account within 28 days of receipt of written notice from the contractor, then the contractor is entitled to a 5 per cent rate of interest on outstanding sums.

The contractor is to remedy, at his own expense, any defects in workmanship or materials occurring before the termination of the maintenance period (clause 40). The site is to be left in good order and free of surplus materials on completion (clause 41).

If the contractor becomes bankrupt or fails to carry out the terms of the contract, the contractor may be expelled from the site and he will not be entitled to any further payment until the end of the maintenance period (clause 42). Clause 43 lists a number of circumstances under which the contractor has the right to cease work and to recover the costs incurred from the employer. Clause 44 makes provision for the settlement of disputes by a single arbitrator where the parties are unable to reach agreement.

TYPES OF CONTRACT

There are basically eight types of contract which can be used for civil engineering work, and each of these is now described and their uses are detailed.

(1) Bill of Quantities Contract

This is generally the soundest form of contract. It incorporates a bill of quantities, in which all the component items of work are listed and subsequently priced by the contractor to give the tender sum. The bill of quantities helps the contractor in pricing the job, and in the absence of this document he may have to prepare his own approximate bill in the limited time allowed for tendering.

Furthermore, the priced bill of quantities provides a basis for the valuation of variations, in addition to ensuring a common basis for tendering. It is good policy to use this form of contract on all but the smallest civil engineering jobs, or where the special circumstances on a job favour the use of some other form of contract.

(2) Lump Sum Contract

In a lump sum fixed price contract the contractor undertakes to carry out the contract works for a fixed sum of money. The details of the project are normally shown on drawings and described in a specification, but no bill of quantities is supplied. This form of contract is often used where the job is reasonably certain in character and small in extent, such as site clearance work and on small constructional jobs like pumping stations.

(3) Schedule Contract

This type of contract may take one of two forms.

(a) The employer supplies a schedule of unit rates covering each item of work likely to be encountered, and contractors are required to state a percentage up or down on the rates, for which they are prepared to undertake the work.

(b) The more usual method is to require contractors to insert a rate against each item of work in the schedule. It is good practice to

12

include approximate quantities to assist contractors in pricing the schedule and for the subsequent comparison of tenders.

This type of contract is used mainly for maintenance work, but occasionally schedules of rates are used as a basis for negotiated contracts.

(4) Cost Plus Percentage Contract

In a cost plus percentage contract the contractor is paid the actual cost of the work, plus an agreed percentage of the actual or allowable cost to cover overheads, profit, etc. It is useful in an emergency, when there is insufficient time available to prepare a detailed scheme before work is begun, but it will be apparent that an unscrupulous contractor could increase his profit by delaying the completion of the works. No incentive exists for the contractor to complete the works as quickly as possible or to try to reduce costs.

(5) Cost Plus Fixed Fee Contract

In this form of contract, the contractor is paid the actual cost of the work plus a fixed lump sum, which has been previously agreed upon and which does not fluctuate with the cost of the job. It is not a particularly good form of contract, although it is superior to the cost plus percentage type contract, as there is some incentive to the contractor to complete the job as speedily as possible and so release his resources for other work.

(6) Cost Plus Fluctuating Fee Contract

In this type of contract the contractor is paid the actual cost of the work plus a fee, with the amount of the fee being determined by reference to the allowable cost of the work on some form of sliding scale. Thus the lower the actual cost of the works, the greater will be the fee that the contractor receives. An incentive thus exists for the contractor to carry out the work as quickly and cheaply as possible, and it constitutes the best of the 'cost plus' or 'cost reimbursement' type of contract, from the employer's viewpoint.

(7) Target Contract

In a target contract a basic fee is quoted as a percentage of an agreed target estimate obtained from a priced bill of quantities. The target estimate may be adjusted for variations in quantity and design and fluctuations in labour and material costs, etc. The actual fee paid to the contractor is determined by increasing or reducing the basic fee by an agreed percentage of the saving or excess between the actual cost and the adjusted target estimate. Thus there is a definite incentive to the contractor to complete the work as speedily and economically as possible, but it involves a considerable amount of contract document preparation, including a bill of quantities.

(8) All-in Contract

With this type of contract the employer gives his requirements in broad outline to the contractor, who submits full details of design, construction and cost for the project. It is suitable in a few special cases, such as gas and chemical works, oil-refineries and nuclear power stations, but it is unlikely to be suitable for the majority of civil engineering contracts (5).

This type of contract is often referred to as a 'package deal'.

REFERENCES

(1) INSTITUTION OF CIVIL ENGINEERS, in conjuction with Association of Consulting Engineers and Federation of Civil Engineering Contractors. *General Conditions of Contract and Forms of Tender, Agreement and Bond for use in connection with Works of Civil Engineering Construction.* Fifth Edition, June 1973.
(2) INSTITUTION OF CIVIL ENGINEERS, in conjunction with Association of Consulting Engineers and Export Group for Constructional Industries, *Conditions of Contract for Overseas Works mainly of Civil Engineering Construction.*
FEDERATION INTERNATIONALE DES INGÉNIEURS CONSEILS. *International Conditions.*
(3) INSTITUTION OF STRUCTURAL ENGINEERS. *General Conditions of Contract for Structural Engineering Works.*
(4) JOINT CONTRACTS TRIBUNAL, *Agreement and Schedule of Conditions of Building Contracts*
    (i) Private edition with quantities;
    (ii) Private edition without quantities;
    (iii) Local authority edition with quantities;
    (iv) Local authority edition without quantities. 1963 (revised July 1973).
(5) INSTITUTION OF CIVIL ENGINEERS. *Civil Engineering Procedure.* 1971.
(6) INSTITUTION OF CIVIL ENGINEERS, *Standard Method of Measurement of Civil Engineering Quantities.* 1953 (reprinted 1972 with metric addendum).

(7) ROYAL INSTITUTION OF CHARTERED SURVEYORS and NATIONAL FEDERATION OF BUILDING TRADES EMPLOYERS, *Standard Method of Measurement of Building Works*. Fifth Edition (Metric) 1968.

(8) DEPARTMENT OF THE ENVIRONMENT, *Specification for Road and Bridge Works*. H.M.S.O., 1969.

(9) SEELEY, Ivor H., *Civil Engineering Quantities*. Macmillan, 1975.

(10) B.S. 308, *Engineering Drawing Practice*.
B.S. 1192, *Building Drawing Practice*.

(11) DEPARTMENT OF THE ENVIRONMENT, *Method of Measurement for Road and Bridge Works*. H.M.S.O., 1971.

(12) FORM GC/WORKS/1: *General Conditions of Government Contracts for Building and Civil Engineering Works*. H.M.S.O., 1973.

(13) INSTITUTION OF CIVIL ENGINEERS, *Civil Engineering Standard Method of Measurement, 1976*.

CHAPTER TWO

Purpose and Arrangement of and Sources of Information for Specifications

THIS chapter is concerned with the functions of the specification, its general arrangement and the sources from which information needed in its compilation can be obtained.

FUNCTIONS OF SPECIFICATION

The specification is a very important contract document on a civil engineering contract, whereas on building jobs it is frequently dispensed with and its place is then taken by lengthy preamble clauses in the bill of quantities. A civil engineering specification is read in conjunction with drawings and a bill of quantities to supply the contractor with all the information he needs to submit a tender and execute the works.

The main function of the specification is to describe fully the workmanship and materials required to carry out the contract. It forms, in fact, a schedule of instructions to the contractor and will, to a large extent, determine the quality of the executed job.

On a civil engineering job, the descriptions of items contained in the bill of quantities are kept as brief as possible, with frequent references to specification clauses. This procedure avoids the duplication of much descriptive information, as full descriptions appear in a single document: the specification. During the execution of the contract, the existence of a comprehensive specification is of great value to the contractor, although at the tendering stage the contractor is obliged to make frequent reference to both documents, which slows down the job of pricing the bill of quantities.

16

In addition to supplying detailed information on the work to be executed and the nature and quality of materials and workmanship, the specification also contains details of any special responsibilities of the contractor which are not covered by the conditions of contract. The specification clauses covering special responsibilities are often termed 'general clauses' as they are not confined to any particular section of work, but relate to the job as a whole. 'General clauses' will be described in some detail in the next chapter.

The specification should, as far as possible, amplify but not repeat and certainly never contradict the information detailed on the contract drawings and given in the bill of quantities and conditions of contract. When the various documents are taken together they should leave no doubt as to the character and extent of the contract works. The specification is a lengthy and complex document and requires considerable skill and wide technical knowledge in its preparation.

General Arrangement of Specifications

Most civil engineering specifications start with 'general clauses' which relate to the job as a whole, and include the special responsibilities of the contractor which are not covered in the conditions of contract. The general clauses are followed by materials and work clauses which are related to the various sections of work making up the job. The latter type of clause is sometimes referred to as a 'special clause'.

The sequence of the materials and work clauses will follow one of two patterns.

(a) Materials clauses are entered first, followed by work clauses in each section (earthwork, concrete work, etc.)
(b) All materials clauses are written first followed by works clauses sub-divided on a sectional basis.

All specification clauses are generally numbered for ease of reference, the reference numbers usually running consecutively throughout the whole of the specification. Each clause is normally provided with a suitable heading, although a clause should be complete in itself without including the heading. Sub-headings act as useful signposts to the contractor. It is advisable to begin each section of a specification on a fresh page, to assist in breaking up the document and making it more readable.

17

DRAFTING OF SPECIFICATION CLAUSES

(1) Material Descriptions

Considerable care must be exercised in the drafting of a specification, to prepare clauses which are concise, complete and free from ambiguity. For instance, when drafting materials clauses it is advisable to adopt some pre-arranged order of grouping the particulars, to avoid missing an important detail. The following build-up of a specification description for bricks will serve to illustrate this approach.

Particulars required	Actual description
Material	Bricks
Type	Southwater red No. 2 engineering bricks
Name of manufacturer or source of supply	Messrs. X of Y.
Prime cost	£55·00 per thousand, delivered to the site.
Desirable characteristics	Well burnt, of uniform shape, size and colour, and sound and hard.
Undesirable characteristics	Free from cracks, stones, lime and other deleterious substances.
Tests	Minimum compressive strength of 48·3 MN/m² (7000 lbf/in.²) Maximum water absorption of 7 per cent by weight.

Note: Reference to the appropriate British Standard will reduce considerably the amount of information to be given.

The following alternative methods of describing materials, or possibly a combination of them, can be used in a specification.

(a) A full description of the material or component is given with details of desirable and undesirable characteristics and appropriate test requirements.

(b) Relevant British Standard reference, together with details of class or type where required is given. The contractor can then refer to the British Standard for fuller information.

(c) Name of manufacturer, proprietary brand or source of supply is stated and the contractor can obtain further particulars from the manufacturer or supplier.

(d) A brief description of the material is given together with the prime cost for supply and delivery of a certain quantity of the material to the site.

Typical quantities are: a thousand bricks, a tonne (ton) of tarmacadam or a cubic metre (yard) of sand, and these normally represent the units by which the materials are sold.

This latter method ensures that all contractors are tendering on the same basis, without the need to obtain quotations from manufacturers or suppliers. It also permits the employer to defer the actual choice of material to a later date if he so wishes.

(2) Avoidance of Unsatisfactory Descriptions

In practice, use is made of a number of wide and embracing terms which are not sufficiently precise in their meaning and can be interpreted in different ways. This leads to inconsistencies in pricing with consequent undesirable effects. Some examples of undesirable terms are now given.

(a) The word 'best' is widely used in specifications, where best quality materials or workmanship are obviously not required. If this term is frequently and loosely applied throughout the specification, without any real consideration as to its true intent and meaning, then the contractor will be tempted to disregard it. It is important to prevent this happening by using the term only when materials or workmanship of the highest quality are required. Materials are frequently produced in a number of grades and it is essential that a clear indication should be given of the particular grade required. For instance, it would be pointless to specify best quality vitrified clay pipes when British Standard surface water pipes were really required.

(b) The word 'proper' is also frequently misapplied, particularly in descriptions of constructional methods. As a general rule it is far more satisfactory to include full instructions in the specification, and so leave the contractor in no doubt as to the actual requirements of the job. With minor items of work a comprehensive description of the method of construction may not be essential and in these circumstances the use of the word 'proper' may be acceptable.

(c) The term 'or other approved' usually represents an undesirable feature in any specification, as it introduces an element of uncertainty. The contractor cannot be sure whether the materials or components which he has in mind will subsequently prove acceptable to the engineer. All specification requirements should be clear and certain in their meaning and be entirely free from doubt or ambiguity.

(d) The term 'as specified' is used widely sometimes without specifying anything.

(3) Workmanship Clauses

Specification clauses covering constructional work and workmanship requirements, are generally drafted in the imperative, e.g. 'Lay manhole bases in concrete, class B, 225 mm (9 in.) thick', or alternatively 'The contractor shall lay, etc.'. All workmanship clauses should give a clear and concise description of the character and extent of the work involved.

The sequence of clauses within a section will normally follow the order of constructional operations on the site. This procedure reduces the possibility of omission of items from the specification and assists the contractor in working to its requirements on the site.

It is essential that the specification clauses should be kept as concise and straightforward as possible, with an absence of unreasonable requirements. Lengthy, involved clauses tend to confuse the contractor and may well result in higher prices. The specification writer must have a sound knowledge of the type of construction which he is specifying and know exactly how it will be executed, in order to be able to draft a specification which is entirely satisfactory in all respects. He must also avoid specifying standards of work and quality of materials which are completely out of keeping with the class of work involved.

The specification is a highly technical document and is drafted in technical language with the free use of recognised civil engineering terms. In this respect, it differs appreciably from a report prepared for an employer which should be kept as free from technical terms as possible.

SOURCES OF INFORMATION

Information for use in civil engineering specifications can be obtained from a variety of sources. The principal sources of information are now described.

(1) Previous Specifications

In the majority of cases specifications for past jobs are used as a basis in the preparation of a new specification for a job of similar type.

This procedure expedites the task of specification writing considerably, but care must be taken to bring the specification clauses up-to-date by the incorporation of the latest developments and techniques. It is also necessary to be constantly on the alert for any changes of specification needed to cope with differences of design, construction or site conditions in the current job. Care must be taken to omit details which are not applicable and to insert information on additional features.

(2) Contract Drawings

The draft contract drawings will be prepared in advance of the specification, and these will show the character and extent of the works. The specification writer will extract a considerable amount of information from the drawings and will work systematically through them when compiling the specification.

(3) Employer's Requirements

The employer or promoter often lays down certain requirements in connection with the job and these will probably need to be incorporated in the specification. Typical requirements of this kind are programming of the works so as to provide for completion of certain sections at specified dates and the taking of various precautions so as to cause the minimum interference with productive processes in the employer's existing premises.

It is essential that requirements of this kind are brought to the notice of the contractor, as they may quite easily result in increased costs.

(4) Site Investigations

Some of the information inserted in a specification will arise from site investigations, such as information on soil conditions and water-table level and the extent of site clearance work. The contractor should be supplied with the fullest information available, to reduce to a minimum the risks that he must take and the number of uncertain factors for which he must make allowance in his tender.

(5) British Standards

Frequent references are made in civil engineering specifications to British Standards relating to materials and components. This practice permits a considerable reduction in the length of clauses relating to materials and components and ensures a good standard of product. It also helps the contractor, as he no longer has to read through lengthy descriptive clauses with wide variations from one contract to another.

(6) Codes of Practice

Reference can also be made in specifications to codes of practice in some instances. This practice ensures a good standard of construction and workmanship without the need for lengthy specification clauses.

(7) Trade Catalogues

Where proprietary articles are being specified for use on a job, reference will be made to the manufacturers' catalogues for the extraction of the necessary particulars for inclusion in the specification. It is often necessary to quote the catalogue reference when an article is produced to a number of different patterns.

This procedure will also reduce the length of specification clauses and will ensure the use of a specific article with which the engineer is familiar and in which he has every confidence. Some public bodies object to this practice as it restricts the contractor's freedom of choice and in some cases prevents the use of local products. Furthermore, it may prevent the contractor from using his regular sources of supply and may thus result in higher prices.

BRITISH STANDARDS

British Standards are prepared by the British Standards Institution, which is the approved body for the preparation and promulgation of national standards covering methods of test; terms, definitions and symbols; standards of quality, of performance or of dimensions; preferred ranges and codes of practice.

The institution has a general council which controls five divisional councils concerned with building, chemicals, engineering, textiles and codes of practice. Over seventy industrial standards committees, each representing a major section of industry, are responsible to the divisional councils. These committees are largely concerned with developing industrial standardisation, as they decide the subject-matter of new standards and their scope and extent, and approve the draft standards which are prepared by various technical committees. The technical committees comprise experts on the subject-matter of each standard and consist of representatives of the users, producers, researchers and other interests.

The enormous scope of the British Standards Institution can be appreciated when it is realised that there are about 5000 operative British Standards. These are kept constantly under review in order that they may be up-to-date and abreast of progress. They have proved to be an efficient means of making the results of research available to industry in a practical form.

British Standards prescribe the recognised standards of quality for a wide range of materials and components, and also define the dimensions and tests to which they must conform. The coverage of British Standards is so extensive that almost every specification contains numerous references to British Standards. Appendix II, at the end of this book, contains a schedule of the principal British Standards relating to civil engineering materials and components.

British Standards are of great value in the drafting of specifications, as they reduce considerably the amount of descriptive work required, and yet at the same time ensure the use of a first-class product. The standards are prepared by committees, each of which is made up of experts in the particular field covered by the standard. Each standard does therefore incorporate the most searching requirements that the latest stage of technical development and knowledge can produce.

The contractor, who will encounter British Standards quite frequently, will possess a reasonable knowledge of their contents and will be freed from the necessity to examine carefully lengthy specification clauses relating to materials and components, in case they introduce some unusual requirement which will affect the price. The use of British Standards in this way materially assists in establishing a uniform basis for tendering, as each contractor is pricing for exactly the same articles.

The specification writer must, however, possess a good knowledge of the scope and contents of any particular standard, prior to making reference to it in a specification. Where different grades or classes of a material or component are given in a British Standard, then the particular grade or class of product required must be indicated in the specification. In practice, one frequently finds references to British Standards in this category without any class or grade being mentioned. Similarly, the specifying of first, second or best quality glazed vitrified clay pipes conforming to B.S. 65:1963 is entirely unsatisfactory, as the only two classes of pipe recognised by this standard are 'British Standard', and 'British Standard Surface Water'.

The British Standards mark (popularly referred to as the 'kite' mark) is a registered certification trade mark owned by the British Standards Institution, which may by licence permit manufacturers to use the mark on their products, after they have agreed to follow a routine of inspection, sampling and testing appropriate to the particular product. The mark is thus an independent assurance to the purchaser that the products are produced and tested in accordance with the requirements of the relevant British Standard and its associated certification scheme.

The British Standards Institution opened a testing and inspection centre at Hemel Hempstead New Town in 1959, and this has since been progressively extended. The functions performed by this centre can be broken down into three main categories.

(a) To control the factory inspections and laboratory testing in connection with the B.S.I. certification marking scheme.
(b) To undertake testing commissions from individual firms and organisations relating to overseas as well as British Standards.
(c) To produce individual schemes of approval to serve the needs of particular industries.

In addition to the publication of British Standards and Codes of Practice, the British Standards Institution issues a British Standards Yearbook, an Annual Report and B.S.I. News. The Yearbook is particularly useful as it contains brief summaries of all operative British Standards and Codes of Practice. Complete sets of British Standards are maintained for reference in a large number of public libraries, universities and technical colleges throughout the United Kingdom.

CODES OF PRACTICE

Codes of Practice are issued by the British Standards Institution and represent a standard of good practice. The principal codes of practice relating to civil engineering work are listed in Appendix I. Codes of Practice cover design, construction and maintenance aspects, are extremely comprehensive in their scope, and are well illustrated. As their contents become more widely known and recognised they should secure improved standards of work within their respective fields of operation.

For instance, Code of Practice CP 2003: Earthworks deals with earthworks ancillary to other civil engineering work, but excludes consideration of tunnels and such works as dams, dykes, canals, dredging and river-training works. The first part of the Code describes methods of carrying out projects for cuttings and bulk excavation in the open and for the construction of embankments on areas of fill in road, railway and airfield construction. The second part covers trenches, pits and shafts, and deals with methods of excavating these in various types of ground and of forming temporary supports to the sides. The Code is intended to form a guide to engineers in the design and execution of works, but it is not intended that it should be used as a standard specification.

As a further illustration of the scope of codes of practice, CP 301: Building Drainage sets out recommendations for the design, layout and construction of foul and surface water drains and sewers not exceeding 300 mm (12 in.) in diameter (together with all ancillary works such as connections, manholes, inspection chambers, etc.), used from the building to the connection to a public sewer or to a treatment works, soakaway or watercourse. It deals with methods of excavation, keeping the work free from water, timbering and supporting, filling and consolidation and surface reinstatement.

CHAPTER THREE

General Clauses in Specifications

MOST specifications for civil engineering work begin with a selection of 'general clauses' which relate to the contract as a whole and are not confined to any particular trade or works section. The nature and content of these clauses will vary from job to job. An attempt is made to classify and group the more commonly used general clauses on a functional basis and it is hoped that this approach will prove useful to the reader.

GENERAL CONTRACTUAL MATTERS

(1) General requirements
(2) General conditions
(3) Extent of contract
(4) Contract drawings
(5) Price variations
(6) Cancellation of Contract owing to offer of gift or reward
(7) Sufficiency of tender

LEGAL PROVISIONS

(1) Acts and regulations
(2) Factories Acts
(3) River Authority regulations
(4) Working rule agreement

General Matters affecting the Cost of the Job

(1) Labour expenses and overtime
(2) Safety precautions
(3) Sequence of works
(4) Electricity supply
(5) Water supply
(6) Contractor to visit site
(7) Access to site
(8) Subsoil investigations

Use of Site

(1) Working area
(2) Use of site
(3) Storage
(4) Advertising

Temporary Works

(1) Diversion of services
(2) Traffic control
(3) Temporary access and works
(4) Dealing with flows in existing pipelines
(5) Interference with existing works
(6) Protection of existing sewers
(7) Surveys and setting out
(8) Office and instruments
(9) Telephone
(10) Sanitary accommodation
(11) Use of roads
(12) Making good damage to existing properties

Materials and Workmanship Requirements

(1) Labour and materials
(2) Plant

27

(3) Materials and workmanship
(4) Samples, tests and certificates
(5) Work prepared off-site

GENERAL WORKING REQUIREMENTS

(1) Suspension of work during bad weather
(2) Facilities for other contractors
(3) Co-ordination with other contractors
(4) Accommodation for employees
(5) Keeping site tidy
(6) Measurement of work
(7) Photographs
(8) Protection of work

RECORDS

(1) Programme and progress record
(2) Other records

To assist readers in the drafting of suitable general clauses for civil engineering work, a large range of typical general clauses follows, accompanied by explanatory notes.

GENERAL MATTERS

TYPICAL GENERAL CLAUSES　　　　EXPLANATORY NOTES

General requirements

The works to be executed under this Contract are to be carried out in accordance with the Conditions of Contract, Specification, Bill of Quantities and Contract Drawings. If in construing the Contract there shall appear to be any inconsistency between the Specification and Conditions of Contract, the Conditions of Contract shall prevail.

This clause draws the Contractor's attention to the requirements of all the Contract Documents and emphasises that the Conditions of Contract have precedence over other contract documents.

General Conditions

The General Conditions of Contract, Forms of Tender, Agreement and Bond applicable to this Contract are those drawn up by the Institution of Civil Engineers jointly with the Association of Consulting Engineers and the Federation of Civil Engineering Contractors (dated June, 1973) and provision for compliance with the clauses in the Conditions of Contract must be made in the measured rates.

Alternatively, the General Conditions of Contract for Structural Engineering Works, issued by the Institution of Structural Engineers, could be operated for structural jobs.

Extent of contract

(a) *Railway bridge over canal.* The Works in this Contract consist of the driving of steel sheet piling alongside the canal; the excavation for and construction of two concrete piers; the fabrication and erection of two welded plate girders with welded diaphragms site welded to the main girders; the supply and placing on the plate girders of precast reinforced concrete trough units; the construction of bored piles in the existing railway embankment and the construction of reinforced concrete abutments; the supply of precast reinforced concrete longitudinal beams and erection as side spans between the abutments and piers and all ancillary works in connection therewith.

(b) *Extensions to sewage disposal works.* The Works in this Contract consist of the construction of a flow-dividing chamber; sedimentation tanks; dosing chambers; biological filters; humus tanks; sludge well and pumphouse; effluent sump and pumphouse, including

Descriptions will vary considerably according to the nature of the work. For this reason two typical clauses are given; one relating to a railway bridge and the other to sewage disposal works extensions.

The approach is rather different in each case, with a much more detailed description being supplied for the railway bridge with information on forms of construction. The sewage disposal works clause merely lists the major component parts of the works.

Basically, the function of this clause is to give the Contractor a general picture of the job before he begins reading the detailed clauses.

pumping plant; sludge drying beds; sludge, top water and effluent pumping mains; feed, effluent and sludge pipes; distribution and collecting grips; manholes; and alterations to existing works.

Drawings

The drawings referred to in the General Conditions of Contract are listed hereunder. Written dimensions should be taken in preference to scaling.

Contract Drawing No.	Title
1	Layout of site roads and surfaced areas
2	Site roads: cross sections, Sheet No. 1
3	Ditto: Sheet No. 2
4	Plan of road drainage
5	Road drainage: Longitudinal sections
6	Road drainage: gully and manhole details

The purpose of this clause is to list the Contract Drawings with their numbers and titles. This permits the Contractor to check that he has all the Contract Drawings in his possession at the tendering stage.

Price variations

The Contract Price Fluctuations Clause, prepared by the Institution of Civil Engineers in conjunction with the Federation of Civil Engineering Contractors and the Association of Consulting Engineers, is to operate on this Contract.

Where this clause is to apply, it must be expressly incorporated in the Contract. Its use is particularly necessary on contracts which are uncertain in extent or of long duration.

Cancellation of contract if contractor shall offer any gift or reward

The Employer shall be entitled to cancel the Contract and to recover from the Contractor the amount of any loss

This clause or a similar one is usually incorporated in a civil engineering contract

resulting from such cancellation, if the Contractor shall have offered or given or agreed to give to any person any gift or consideration of any kind, as an inducement or reward for doing or forbearing to do or for having done or forborne to do any action in relation to the obtaining or execution of the Contract, or any other Contract with the Employer or for showing or forbearing to show favour or disfavour to any person in relation to the Contract or any other Contract with the Employer, or if the like acts shall have been done by any person employed by him or acting on his behalf (whether with or without the knowledge of the Contractor) or if in relation to any Contract with the Employer the Contractor or any person employed by him or acting on his behalf shall have committed any offence under the Prevention of Corruption Acts 1889 to 1916, or shall have given any fee or reward the receipt of which is an offence under sub-section (2) of section 123 of the Local Government Act 1933.

for a local authority with the object of preventing any bribery or corruption occurring on such a contract.

Sufficiency of tender

The Contractor shall be deemed to have satisfied himself before tendering as to the correctness and sufficiency of his tender for the Works and of the prices inserted in the Bill of Quantities (and Schedule of Rates, where applicable), which rates and prices shall cover all his obligations under the Contract and everything necessary for the proper completion of the works.

This clause operates as a reminder to the Contractor that his tender must include all the works (both permanent and temporary) and general obligations and liabilities involved in the completion of the Contract.

LEGAL PROVISIONS

Acts and regulations

The Contractor shall comply with all Acts of Parliament and Statutory Instruments applicable to the Contract Works and shall indemnify and keep indemnified the Employer against any failure on the part of the Contractor or his employees to comply with any such Acts of Parliament or Statutory Instruments and against any damages or other consequences arising out of such failure. In particular, the Contractor's attention is drawn to The Building (Safety, Health and Welfare) Regulations, 1948 (S.I. 1948 No. 1145), to the Factories Act, 1961, The Health and Safety at Work Act, 1974 and to all subsequent amending legislation and relevant Statutory Instruments.

It is advisable to draw the Contractor's attention to his obligations resulting from Acts of Parliament and Statutory Instruments.

Port or River Authority regulations

The Contractor is to comply with the Port Authority Regulations insofar as they control and regulate any of the Works or the incidental movement of plant, equipment, labour, materials or craft conveying these in or about the Works or plying between the Works and the Contractor's supplying wharves. The Contractor shall obtain from the Port Authority such permissions as are required for carrying out the work or any necessary temporary works. He shall provide, maintain and remove, when required, such fenders, lights, fences, hoardings, guard rails, etc., as may be required by the Port Authority's regulations or on the direction of the

Where the Works adjoin a harbour or river, it is often necessary for the Contractor to comply with the regulations of the Port or River Authority. These regulations may have considerable influence on the Contractor in his execution of the Work and he will need to make allowance for this in his tender.

Engineer. The tender will be deemed to include for the whole of the cost of these operations. On completion of the Works the Contractor shall remove all temporary works and leave the river bed free from any obstructions or deposits arising from the Works.

Working rule agreement

The Contractor shall allow in his Tender for all costs incurred in complying with the provisions of the Working Rule Agreement made by the Civil Engineering Construction Conciliation Board for Great Britain.

This clause emphasises the fact that no additional payments will be made to the Contractor in connection with the working rule agreement.

MATTERS AFFECTING COST

Labour expenses

Subject to the Price Variation Clause, the tender shall include for all expenses relating to labour and in particular the following: insurance of all kinds, selective employment tax, pensions, holidays with pay, overtime working, night work, double shifts, rotary shifts, tidework, tool allowances, servicing of plant, 'plus' rates and allowances and payment by result or bonus.

The Contractor must not merely build-up his billed rates from basic labour rates but must include for a wide range of indirect labour charges.

Safety precautions

The Contractor shall take all necessary precautions and shall comply with all regulations and the recommendations of Reports of the Committee of the Institution of Civil Engineers, dealing with shaft and tunnel works, and for

The wording of this clause will vary with the nature of the work in the Contract. For instance, a sewage works contract would make reference to the publication issued

33

work carried out in compressed air, inside cofferdams, on or below water, and shall continually operate all safety measures made necessary by virtue of the method employed for the execution of the Contract.

by the Institution of Civil Engineers, *Safety in Sewers and at Sewage Works*, whereas with works of water supply the appropriate report is *Safety in Wells and Boreholes.*

Sequence of works

The works are to be carried out in accordance with a programme and in a sequence to be approved by the Engineer in relation to other works to be executed on the site under other Contracts.

The Engineer may from time to time, by order in writing without in any way vitiating the Contract, require the Contractor to proceed with the execution of the Work at such time or times as may be deemed desirable, and the Contractor shall not proceed with any work ordered to be suspended until he receives a written order to do so from the Engineer.

This clause sometimes includes a detailed schedule giving the dates by which specific sections of the Works are to be completed. On occasions, the Contractor is required to make allowance in his tender for possible delays resulting from programme changes introduced by the Engineer: this is an unreasonable request on account of the uncertainties involved.

Electricity supply

The Employer will make available from points on the site as shown on the Site Plan:
(1) a supply of electrical energy of stated kVA at 415 volts, three-phase, for supplying power for heavy erection plant and welding;
(2) a supply of electrical energy at 110 volts, single-phase or three-phase, with centre points earthed, for supplying power to portable tools and for fixed lighting and handlamp transformers.

This clause gives details of the electricity supply available on the site and lists the requirements relating to the Contractor's electrical installation.

The Contractor will be responsible for providing and maintaining the whole of the installation on the load side of the points of supply. All necessary safety precautions must be taken and the Contractor's electrical installation must be to the satisfaction of the Engineer, and must comply with all appropriate statutory requirements including the current Regulations for the Electrical Equipment of Buildings issued by the Institution of Electrical Engineers.

Water supply

The Contractor will be supplied free of charge at a fixed point on the site with the necessary water for constructional purposes. The Contractor shall distribute at his own expense to all other points on the site where water is required, and he will be held responsible for any unnecessary wastage of water and shall be charged for such wastage at current rates.

The Contractor will be held responsible for compliance with the requirements of the local water undertaking.

Alternatively, the Contractor may be required to obtain the necessary supply of water and to pay the water authority's charges. These charges may be calculated as a percentage of the contract sum, although this procedure offers no incentive to the Contractor to prevent the misuse of water.

Contractor to visit site

The Contractor is requested to visit the site of the proposed works before tendering and will be deemed to have satisfied himself as to local conditions, the nature and degree of accessibility of the site, the nature and extent of the operations, the supply of and conditions affecting labour and materials and the execution of the Works generally. No claim will be entertained in respect of

This clause serves to emphasise the need for the Contractor to make a thorough examination of the site prior to submitting his tender, and to make allowance in his price for any factors which will affect working conditions on the site.

any of these matters, neither will lack of knowledge nor ignorance of conditions be accepted as substantiating a claim.

Access to site

Access to the site is obtained from Road X. The Contractor shall provide and maintain such temporary roads as he may require for the purpose of carrying out the work in the most expeditious and efficient manner, and shall remove the temporary roads on completion. Temporary roads shall be constructed of hardcore, engine ash, timber sleepers or other suitable material. The Contractor shall comply with all police and highway authority requirements.

This covers access to the site and the construction, maintenance and subsequent removal of temporary roads on the site, and compliance with all public authority requirements.

Subsoil investigations

Borings have been taken on the site and the results are shown on Drawing X, but there is no guarantee that the conditions found in the borings are truly representative of conditions generally on the site. The Contractor is therefore advised to make his own independent enquiries and observations as to the character of the soil.

All levels on the Contract Drawings are related to Ordnance Datum (Newlyn). Details of subsoil water levels in the ground are also recorded on Drawing X.

The results of trial borings should always be made available to Contractors tendering for a job.

In the case of work to be carried out beside the sea or tidal waters, a number of water levels should be supplied, such as highest recorded water level; mean high water spring tide; mean low water spring tide; and lowest low water.

36

USE OF SITE

Working area

The area available for the storage of materials and assembly of components is indicated on Drawing *X*.

This provision is particularly important when working space is restricted.

Use of site

The Contractor shall, except when authorised by the Engineer, confine his men, materials and plant within the site of which he is given possession. The Contractor shall not use any part of the site for any purposes not connected with the Works unless the prior written consent of the Engineer has been obtained.

The purpose of this clause is to confine the Contractor's activities to the Site of the Works, and to prevent its use for improper purposes. Reference may also be made to the erection of fences and hoardings around the site.

Storage

The Contractor shall provide at his own expense suitable offices and adequate storage accommodation for plant and materials. In particular, adequate waterproof storage sheds shall be provided for materials requiring protection against weather, humidity or damage. No materials or plant shall be stored on the public highway.

It is important that the Contractor should be required to provide adequate and suitable storage accommodation to prevent damage or deterioration to materials and plant.

Advertising

The Contractor shall treat the Contract and everything within it as private and confidential. In particular, the Contractor shall not publish any information, drawing or photograph relating to the Works and shall not use the site for advertising purposes, except with

This requires the Contractor to obtain the Engineer's consent before he can use anything connected with the Contract or even its site for advertising purposes.

37

the written consent of the Engineer and subject to such conditions as he may prescribe.

TEMPORARY WORKS

Diversion of services

Where it becomes necessary permanently to divert any cable, sewer, drain, main, etc., the Contractor will be permitted to recover the cost of this work unless, in the opinion of the Engineer, it was due to the Contractor's negligence, bad workmanship, faulty materials or lack of reasonable foresight. The Contractor is required to make all necessary arrangements with the appropriate authorities for the diversion of their services.

The Contractor is assured of payment for the unavoidable diversion of services, but is responsible for making the necessary arrangements with the authorities concerned.

Traffic control

The Contractor shall provide all necessary traffic control signs and signals that may be required by the Highway Authority or the Police, and shall operate and maintain them efficiently and to the satisfaction of the authorities mentioned. The contract rates for pipelaying shall be deemed to include all expenses of traffic control, watching and lighting.

This provision is particularly appropriate on sewer and pipeline contracts, where work is to be carried out along public highways.

Temporary works

The Contractor shall be solely responsible for the sufficiency, stability and safety of all temporary works and their care in accordance with clause 20 of the

This clause gives added emphasis to clause 20 of the I.C.E. Conditions and provides for the supply to the

I.C.E. Conditions. He shall at his own expense supply detailed drawings and calculations of stability of such temporary works as the Engineer may direct, but no approval given or implied by the Engineer shall relieve the Contractor of his responsibilities in connection with temporary works.

Engineer of drawings and calculations relating to temporary works.

Dealing with flow in existing pipelines

The Contractor shall at his own expense deal with, pump, maintain or divert the flows in existing pipelines as may be necessary during the execution of the works.

This provision may be necessary when the Contract Works involve alterations to existing operative sewers and pipelines.

Existing services

The Contractor shall avoid all damage to operative gas or water mains or service pipes, sewers, drains, cables, wires, overhead telephone or telegraph lines, telegraph poles, etc., which he encounters in carrying out the Works and he must provide for supporting them to the satisfaction of the Engineer and the responsible authorities. The Contractor must give all notices, pay any charges made and make good at his own expense any damage done. The Contractor must satisfy himself by his own enquiries and observations as to the precise position of all existing services and must take full responsibility in connection with them and shall hold the Employer indemnified against any claims that may arise from any damage caused to such services by his operations.

This is an important clause, as costly damage is often caused to electricity cables and other services by plant engaged on excavating work. It is essential that the Contractor takes full responsibility for all damage to and works of support required in connection with existing services, otherwise the Employer could be liable for the cost of work resulting from the negligence of the Contractor.

The Engineer normally gives an indication of the extent of existing services so far as they are known to him, but as a guide only since their true whereabouts can only be determined by the Contractor.

Setting out

The Contractor shall be entirely responsible for accurately setting out all the work, and he shall at his own expense make good any defects arising from errors in the lines or levels. He shall also provide for the use by the Resident Engineer throughout the contract, a modern and accurate theodolite and precision level, both of approved type and make, complete with all ancillary equipment, steel and linen tapes, poles, pegs, stagings, templates, profiles, etc., necessary for setting out and measurement of the work, and the services of an experienced chainman. The Contractor shall also provide such rubber boots, oilskins and protective clothing as may be required by the Engineer's staff on the site.

Although the Contractor is responsible for setting out the work the Engineer or his representative normally carries out checks and needs a supply of instruments, equipment and a chainman for this purpose.

Office for Resident Engineer

The Contractor shall provide and erect in a position approved by the Engineer, and maintain, clean, heat and light throughout the Contract, one suitable and substantial office measuring not less than 4·5 m × 3·75 m (15 ft × 12 ft) internally, with a close-boarded floor and adequate windows, for the sole use of the Resident Engineer and his staff. The office shall contain one locked cupboard, one table 2·5 m (8 ft) long with two locked drawers and three double elephant plan drawers, one office desk, one typewriter, two chairs, two stools, one double elephant drawing board and T-square, and washing facilities, and it shall be provided with a

The office requirements will vary from job to job according to the size of a contract and the number of engineer's staff employed on the site. It is necessary to give detailed particulars of the office and its contents in order that the contractor can make a realistic assessment of their probable cost at the time of tendering.

Site offices often have to be large enough to accommodate site meetings.

lock and four keys. There shall be no spare keys in the possession of any person other than the Engineer or his representative.

The Contractor shall provide such labour as is reasonably necessary to attend to the office requirements, clean instruments and assist in measuring, supervising, checking or testing the work at any time.

Telephone

The Contractor shall arrange for the installation of a Post Office telephone in the Resident Engineer's office, which is to be kept directly connected to the public telephone exchange, and the Contractor shall pay all charges for installation, rent, calls and eventual disconnection. The Employer will reimburse to the Contractor the sums paid to the Post Office in this connection. The Contractor will not be permitted to have any other telephone on the same line or as an extension from it.

The Engineer's representative normally requires the sole use of a telephone on the site, and a provisional sum is generally included in the bill of quantities to cover the costs involved, since the Contractor cannot be reasonably expected to estimate the probable cost of telephone calls to be made from the Engineer's office.

Sanitary conveniences

The Contractor shall provide and maintain at his own expense proper and adequate sanitary conveniences for the use of persons on the site throughout the Contract, and shall remove same on completion. Where water closets cannot be provided, the Contractor shall provide suitable chemical closets which shall receive regular attention. All sanitary accommodation shall be to the approval of the Engineer and the local Public Health Authority.

This clause supplements the requirements of the Safety, Health and Welfare Regulations.

Use of public highways

The Contractor shall take all reasonable precautions to prevent the deposit of mud, filth or rubbish on the highway and shall from time to time, or as instructed by the Engineer or the Highway Authority, remove from the highway at his own expense any mud, filth or rubbish which may have been deposited on it.

This clause gives added emphasis to the Contractor's responsibility to keep public highways free from mud, etc., from the site. Failure of the Contractor to keep roads clean can cause considerable inconvenience and annoyance to the public.

Damage to adjoining properties

The Contractor shall not trespass on properties adjoining the site of the Works. The Contractor will be held responsible for any damage to adjoining properties caused through the carrying out of work contained in this Contract. The Contractor shall repair and make good any such damage at his own expense to the satisfaction of the Owners.

It is important that the Employer should be indemnified against any claims submitted by adjoining owners.

MATERIALS REQUIREMENTS

Plant, etc.

The Contractor is to provide all labour, materials, plant, tools, tackle, etc., necessary for the satisfactory completion of the Works. All mechanical plant used by the Contractor in the execution of the Works shall be of such type and size and subject to such method of working as the Engineer may approve.

The Contractor must provide everything necessary for the job and it is desirable that the Engineer should have some control over the type of plant and how it is used.

Materials and workmanship

All materials and components shall be of good quality, appropriate to the class of work involved, and be in full accordance with the Contract requirements. Where an applicable specification issued by the British Standards Institution is currently operating, the materials and components used in the execution of the Works shall comply with that specification, unless otherwise specified or ordered by the Engineer.

Workmanship shall be of a high standard and shall conform to the detailed requirements of the specification and the appropriate sections of any applicable current Codes of Practice issued by the British Standards Institution.

It is unsatisfactory to require all materials and components to be of the best quality, as this is rarely always the case. It is good and sound practice to make wide use of British Standards and Codes of Practice.

Samples, tests and certificates

When required by the Engineer, the Contractor shall at his own expense submit to the Engineer, for approval, samples of any of the materials and components to be used. The quality of materials and components subsequently used in the Works shall not be inferior to the approved samples.

The cost of testing is to be borne by the Contractor, who shall give not less than 7 days notice of all tests in order that the Engineer or his representative may be present. Two copies of all test certificates shall be supplied to the Engineer or his representative.

All material which is specified to be tested at the Manufacturer's Works must satisfactorily pass the tests before being painted or otherwise covered.

To assist the Contractor and to clarify the testing procedure, it is the practice of some Engineers to detail the number of certificates required for the principal materials and components. For instance, one test certificate may be required for each delivery of cement, and each type of pipe, one for each sluice valve, one for each 10 tonnes (10 tons) of bar reinforcement of each diameter, one for each make of brick, etc.

43

Test certificates shall be supplied to the Engineer or his representative before the materials or components are used in the Works, unless the Engineer directs otherwise.

Work prepared off the site

The Contractor shall give the Engineer written notice of the preparation or manufacture at a place off the site of any material or component to be used on the Works, stating the place and time of preparation or manufacture, so that the Engineer may make inspections at all stages of the production process. Failure to give such notice may result in the rejection of the material or component, if the Engineer considers that his inspection was necessary during the production process.

The Engineer needs the facility to inspect the preparation or manufacture of materials and components away from the site.

GENERAL WORKING REQUIREMENTS

Suspension of works during bad weather

The Contractor shall, without compensation, delay or suspend the progress of the Works, or any part thereof, during frost or bad weather for such periods as may be required by the Engineer. The Engineer shall determine what extension of time (if any) shall be allowed to the Contractor for such suspensions.

This clause empowers the Engineer to order the suspension of any part of the Works during exceptionally bad weather, without incurring the Employer in additional expenditure.

Facilities for other contractors

The Contractor shall afford all reasonable facilities to other Contractors employed by the Employer, or to any

On some civil engineering contracts, notably those for power stations, it is necessary

local or other authority, to execute work on the site. The Contractor will not be held responsible for injury to such other work or workmen employed on it, unless the injury is caused by the Contractor's operations or by persons in his employ.

for several main contractors to be engaged on the site at the same time, e.g. civil, mechanical and electrical contractors. It is not then possible for a single contractor to have exclusive possession of the site.

Co-ordination with other contractors

The Contractor shall co-ordinate his work with that of the other contractors, so as to cause the minimum practical interference with their work. The other contractors will likewise be required to enter into reciprocal arrangements. The Contractor shall bear all reasonable costs or charges that, in the opinion of the Engineer, are caused by the Contractor's lack of reasonable co-operation.

This provision is frequently necessary on civil engineering contracts, when the Employer also enters into separate contracts for such work as piling and installation of machinery.

Accommodation for employees

The Contractor shall allow in his tender for providing all necessary canteen and first-aid facilities and other accommodation and services for his employees, and shall provide for maintaining them in a clean and tidy condition throughout the construction period, and for clearing away and reinstating the site on completion, all to the satisfaction of the Engineer and other appropriate authorities.

The Contractor has to allow in his price for all canteen, first-aid and other welfare facilities needed for his employees on the site.

Keep site tidy

The Contractor shall throughout the constructional period maintain the whole of the site and all plant and

This clause provides for keeping the site in a tidy condition and for cleaning

materials placed on it in a clean, tidy and safe condition to the satisfaction of the Engineer. The Contractor shall clear away all rubbish from time to time as directed and at completion.

The Contractor shall clean down the surfaces of all concrete, cladding and other work from time to time, and wash pavings and flush drains and gullies. He shall allow for cleaning down the whole of the Works at completion and leaving them in a clean and perfect condition to the satisfaction of the Engineer.

down walls and pavings, flushing out drains, etc.

Measurement of work

The Contractor shall provide a suitably qualified agent and chainmen to assist the Engineer/Quantity Surveyor who will be responsible for measurements, interim valuations and measurement for the final account. The measurements and the form in which the accounts are submitted by the Contractor shall be in accordance with the reasonable demands of the Engineer/Quantity Surveyor.

Whenever the Contractor shall carry out any work or provide any material for which he may propose to claim an extra, he shall first obtain a written order from the Engineer and then make arrangements for its joint measurement, although this will in no way commit the Engineer to recognition of the claim. The Engineer/Quantity Surveyor shall at all reasonable times have access to the Contractor's time-book.

Measurement of work on a civil engineering contract is normally carried out by the Engineer's representative in conjunction with the Contractor's agent. On occasions a quantity surveyor performs the Engineer's measuring functions.

This clause emphasises the need for joint measurement, and outlines the procedure for dealing with 'extra work'.

Photographs

A provisional sum of £150 (One hundred and fifty pounds) is included

Photographs are often required on civil engineering

in the Bill of Quantities for Progress Photographs. This sum is to be expended in whole or in part as directed by the Engineer, or deducted if not required.

jobs to provide a permanent pictorial record at various stages of the job. These often assist in assessing the extent of delays and sometimes help in settling disputes when measuring work. As the Contractor cannot possibly estimate the number of photographs that will be required, the only satisfactory way of dealing with this in the Bill is by means of a provisional sum.

Protection of work

The Contractor shall, at his own expense, cover up and protect all materials and work liable to be stained or injured from any cause, and shall make good any such damage to the entire satisfaction of the Engineer. The Contractor shall adequately protect from frost all concrete, brickwork, masonry, rendering and other work requiring the use of cement, and shall not execute such work when the temperature of the atmosphere or materials is below 2°C (36°F), except with the Engineer's consent and subject to any additional precautions that he may prescribe. Adequate protection shall also be provided against hot sun or rain.

The Contractor is made responsible for protecting all constructional work from damage from any cause, including the effects of extreme weather conditions.

RECORDS

Programme and progress record

The programme which the Contractor is required to provide under clause 14 of the I.C.E. Conditions shall be in a

It is essential that proper provision is made for satisfactory programming and

form approved by the Engineer. The Contractor shall from time to time modify the programme if required to do so by the Engineer. The Contractor shall at all times during the progress of the Works endeavour to adhere to the approved programme.

The Contractor shall supply to the Engineer weekly during the progress of the Works such written particulars and information as will enable the Engineer to maintain a progress record for the Works in the same form as the approved programme.

progressing. This requires the submission of adequate information weekly by the Contractor to the Engineer.

Records

The Contractor shall maintain accurate records, plans and charts showing the dates and progress of all main operations, and the Engineer shall have access to this information at all reasonable times. Records of tests made shall be handed to the Engineer's representative at the end of each day.

The Contractor shall also maintain records and charts of all strata and materials encountered in shaft sinking and tunnel driving, together with records of working conditions under compressed air. The Engineer shall be supplied with a copy of these records as and when required.

The Contractor is normally required to maintain records of progress made, tests performed, strata encountered and details of working conditions, and to submit this information to the Engineer as it becomes available.

CHAPTER FOUR

Specification of Earthwork

THE majority of civil engineering projects entail a considerable amount of earthwork. Apart from general excavation, filling and the disposal of excavated material, some jobs require the execution of more specialised forms of excavation work, such as tunnel work and dredging. In addition it is customary to include, under the general heading of 'Earthwork', specification requirements covering ancillary works such as timbering and keeping excavations free from water.

It will probably be useful at this stage to consider the matters that may require inclusion in an earthworks specification and to determine a logical sequence for them. The specification writer can then determine from this list the items which he needs to incorporate in the specification for his particular job.

PRELIMINARY WORK

(1) Site investigation
(2) Site clearance

EXCAVATION, FILL AND DISPOSAL

(1) Excavation work generally
(2) Excavation of pipe trenches
(3) Excess excavation
(4) Disposal of surplus excavated material
(5) Backfilling
(6) Fill
(7) Trimming slopes
(8) Restricted use of plant

49

ANCILLARY WORK

(1) Keeping excavations free from water
(2) Timbering

SPECIALISED WORK

(1) Tunnel work
(2) Cofferdams
(3) Dredging

The Codes of Practice on earthworks (CP 2003) and site investigations (CP 2001) contain a vast amount of detailed information on the methods which can be used in carrying out these classes of work. They can be useful guides to the Engineer in the drafting of specifications.

A wide selection of typical earthwork specification clauses follows, accompanied by explanatory notes, although it will be appreciated that the detailed requirements will vary considerably from job to job.

TYPICAL SPECIFICATION CLAUSES   EXPLANATORY NOTES

PRELIMINARY WORK

Site investigation

Borings and ground information. Drawings showing details of the sub-strata obtained from borings are available for inspection at the offices of the Engineer. The Contractor is requested to examine these drawings and to make himself fully familiar with the conditions on site.

The information on these drawings is believed to be correct but is not guaranteed and is supplied for guidance purposes. The Contractor is responsible for obtaining such additional information as he considers necessary covering such matters as the nature of the ground, water levels, physical features of the site, etc.

It is customary to supply contractors with information on sub-strata and water-tables from borings, subject to the proviso that these details may not truly represent conditions over the whole area of the site. The onus is on the Contractor to obtain such further information as he deems necessary.

50

Site levels. All levels shown on the Contract Drawings or mentioned in this document are related to Ordnance Datum at Newlyn.

It is the usual practice to relate all levels to Ordnance Datum, preferably a specific bench mark near the site.

Water levels. It is believed that the following truly represent the operative water levels, but their accuracy is not guaranteed.

Highest recorded water level
$$+15\cdot270 \text{ O.D.}$$
Mean high water spring tide
$$+10\cdot310 \text{ O.D.}$$
Mean low water spring tide
$$-\ 8\cdot680 \text{ O.D.}$$
Lowest low water $-12\cdot850$ O.D.

With tidal work it is desirable that the Contractor should be supplied with the various water levels.

Trial holes. The Contractor shall excavate all trial holes as required ahead of pipelaying work, and shall backfill and reinstate them and maintain the surfaces. The Contractor shall receive separate payment for this work provided he obtains prior consent of the Engineer, but he shall at his own expense take all other practicable measures to determine the location of other services.

It is often necessary to determine accurately the location of other underground services before the lines of new sewers and mains can be established with certainty. It is not always possible to obtain sufficiently precise information from the statutory undertakers themselves.

Site clearance

Removal of hedges and brushwood. The Contractor shall uproot and burn all hedges and brushwood on the site of the Works.

These need to be uprooted, and burning on the site is usually the best method of disposal.

Removal of trees and tree stumps. Where indicated on the Drawings or directed by the Engineer, trees shall be uprooted or cut down close to ground level and removed from the site. Stumps and tree

Here again, the removal of roots is essential. Entire trees can often be removed by a tree puller or a bulldozer (see Ministry of Agriculture pam-

roots shall be grubbed up or blasted and burnt or removed from the site. Where directed by the Engineer the voids resulting from tree root removal shall be filled with approved excavated material in 225 mm (9 in.) consolidated layers.

Demolition of concrete wall. Demolish the concrete wall as indicated on the Drawings and retain the resulting debris on the site for use as filling under roads.

phlet No. 101). Alternatively, trees should be cut off at a height not exceeding 1 m (3 ft) above ground and the stumps removed by winch gear or high-velocity explosives (subject to Home Office Regulations and local police requirements) or a combination of both methods.

With demolition work it is important to indicate whether the materials resulting from the demolition are to become the property of the Contractor or the Employer.

Excavation, Fill and Disposal

General

Surface soil. Surface soil shall be stripped and deposited in temporary storage heaps, preparatory to being used for the soiling of slopes to cuttings and embankments and in the preparation of beds to receive trees and shrubs.

Where surface soil is to be retained, provision must be made for its stripping as a separate item from general excavation.

Turf. Where directed by the Engineer, turf shall be carefully cut 1 m × 300 mm × 40 mm (3 ft × 1 ft × 1½ in.) thick and it must be relaid within one week of cutting.

Turf is preferable to soiling and seeding in areas subject to constant and heavy wear, or where a grassed surface is required urgently.

Excavation

The excavation shall be carried out to the dimensions, levels, lines and profiles indicated on the Contract Drawings or

The Contractor is invariably required to excavate to the required dimensions in to such other dimensions as may be directed in writing by the Engineer. The excavations are to be performed in whatever material may be found, but extra payment will be made for excavation in rock. Rock is defined as material which, in the opinion of the Engineer, can be removed only by the use of compressed-air plant, wedges or explosives and is found in continuous beds exceeding 75 mm (3 in.) in thickness or in isolated stones exceeding 0·06 m³ (2 ft³) in volume.

The Contractor shall include for getting out the excavated material by hand or machine and for levelling and ramming surfaces prior to commencing any constructional work.

The faces and beds of all excavations, after being excavated to the required dimensions, shall be carefully trimmed to the required profiles and levels and cleaned of all loose mud, dirt and other debris. The bottom 150 mm (6 in.) of material in the beds of excavations shall not be removed until immediately prior to the execution of the permanent work.

Should slips of material occur during the execution of the Works or during the maintenance period, the Contractor shall be required to perform the remedial works at his own expense, unless the Engineer certifies that the slip occurred through no fault of the Contractor.

No excavation shall be refilled or built upon until the formation has been inspected and approved by the Engineer. Where the Contractor excavates below the required level in good ground, he shall make up the void in concrete (1:3:6) at his own expense.

whatever material may be encountered. This clause will also contain any requirements as to the method of working, such as excavating to within 150 mm (6 in.) of formation in the first instance and removing the latter portion just before construction of the permanent works is commenced.

The choice of method of excavation is usually left to the Contractor and is influenced by a number of factors:
(1) type of ground to be excavated;
(2) ground-water conditions;
(3) the period and height for which an excavated face will stand unsupported;
(4) dimensions of trench.

Mechanical excavation of trenches is normally practicable to a depth of about 3 m (10 ft) in medium ground and 6 m (20 ft) in good ground, using various types of trencher. In wet or bad ground, it is necessary to install an efficient de-watering system before mechanical excavation proceeds.

Excavation of pipe trenches

Pipe trenches shall be excavated to the lines and levels shown on the Contract Drawings or as directed by the Engineer. The bottoms of all trenches shall be of such a width as to provide at least 150 mm (6 in.) clearance between the barrels of the pipes and the excavation or timbering, or such greater clearance as may be necessary to accommodate the concrete surround of the thickness prescribed.

This clause aims at securing pipe trenches of adequate width to allow for satisfactorily laying and jointing the pipes.

Disposal of surplus excavated material

All surplus excavated material shall be removed from the site to a tip to be provided by the Contractor. The site shall be kept as free as possible from accumulations of surplus material.

The responsibility for finding a method of disposing of surplus excavated material almost invariably falls upon the Contractor.

Backfilling

Backfilling of trenches and other excavations shall be carried out in layers not exceeding 225 mm (9 in.) in thickness and the filling shall be of material selected from the excavations and approved by the Engineer. This material must be carefully placed around foundations, etc., and be thoroughly consolidated, with care being taken to prevent any damage being caused to the permanent works. No backfilling shall be deposited until all silt, mud or other soft material has been removed.

In backfilling pipe trenches special care shall be taken to ram the fill at the sides of pipes. The filling material used

This clause aims at ensuring backfilling with suitable materials which are adequately consolidated to eliminate the source of much future trouble and expense. With pipe trenches, special precautions must be taken to prevent any possibility of damage to pipes during ramming of backfilled material.

The degree of compaction of backfilled material will depend on the nature of the work, i.e. on how much soil movement can be tolerated.

54

beside pipes and for 300 mm (12 in.) above them shall be free from stones exceeding 25 mm (1 in.) gauge. The first ramming shall take place at half the height of the pipe. Where a mechanical rammer is used, the pipes shall be protected by at least 1 m (3 ft) of hand-rammed material.

An alternative approach here and elsewhere is to provide a performance specification, prescribing the quality of the end product. For example, fill material to be compacted to an air void content not exceeding 10 per cent.

Fill

Selected and approved excavated material may be used for fill, but certain additional material may have to be imported. Suitable imported material consists of stone, rubble, broken brick and slag, free from objectionable rubbish or trade waste and to the approval of the Engineer. Sample loads of filling material shall be approved by the Engineer before it is used in the Works. Any filling materials rejected by the Engineer shall be removed from the site immediately at the Contractor's expense.

General filling to embankments, etc., shall be thoroughly consolidated with a roller or crawler tractor weighing not less than 10 tonnes (9·84 tons). The fill is to be consolidated in layers not exceeding 300 mm (12 in.) consolidated thickness. Each layer shall extend over the full width of the embankment, and shall consist of reasonably well graded material, with all large voids suitably filled before the next layer is deposited.

The Contractor shall reconstruct to the proper level and profile any filled areas which may settle or spread during the execution of the work or the maintenance period. Where the settlement or

Provision is made for the use of suitable material, adequately consolidated in layers in fill.

Code of Practice CP 2003: Earthworks gives guidance on the design and construction of embankments. Where embankments are to cross soft ground, undisturbed samples should be taken from the foundation at various depths and tested in a consolidation press. From the results of these tests an estimate can be obtained of the settlement that will occur under the fill at various depths.

Where the required height of bank is such that sheer failure of the foundation material might occur, it is advisable to construct the bank in stages, allowing sufficient time for the foundation material to consolidate under the weight of the first stage and so increase its strength before the second stage is added.

spreading is, in the opinion of the Engineer, due to a cause within the control of the Contractor, then the cost of the remedial work shall be borne by the Contractor.

Trimming of side slopes

Side slopes shall be trimmed evenly to the inclinations shown on the drawings or to such other inclinations as the Engineer may direct. Earth slopes and verges, after trimming, shall be soiled to a depth of at least 150 mm (6 in.) with suitable vegetable soil, which shall be sown in the correct season with grass seed as specified at the rate of 0.05 kg/m^2 ($1\frac{1}{2}$ oz/yd^2).

The contract rates for grass seed are to include for preparing and re-sowing bare patches either at the end of the maintenance period or within six months of sowing the seed, whichever is the longer, and for applying a suitable selective weed killer when the grass has become established.

In addition drainage may be required
(1) in the bottom of cuttings;
(2) upon the slopes to intercept springs or seepage water;
(3) at the top of slopes to intercept surface water, either in the form of open channels or French drains. CP 2003: Earthworks contains a useful table covering the design of slopes in various types of rock.

Restricted use of plant

If for any reason the Engineer is of the opinion that it is undesirable that any excavator, mechanical digger or other plant used or proposed to be used by the Contractor for the purpose of excavation should be used or that any such plant is unsuitable for use on the Works or any part of them, the Engineer may order the Contractor not to use and/or to remove the plant from the site.

Although the Engineer does not normally direct the Contractor as to the method in which he carries out the Contract Works or as to the plant which he shall use, nevertheless there may be a case on occasions for restricting the type of plant used to avoid undue annoyance or nuisance being caused to adjoining owners/occupiers.

56

ANCILLARY WORK

Keeping excavations free from water

All excavations shall be kept free from water at all times and adequate pumping plant, including special de-watering equipment, shall be provided by the Contractor, who shall also make his own arrangements for the disposal of all water encountered in the excavations. All sumps shall be located clear of excavations for permanent work, and when no longer required the sumps shall be filled in with suitable material or dealt with as directed by the Engineer.

The Contractor will not be permitted to carry out any concreting or other constructional work unless the excavations are dry, and the excavations shall be kept free from water until the concrete has set sufficiently so as not to be damaged by water.

The Contractor is responsible for keeping all excavations free from water. Ground-water can often be excluded by surrounding the excavations with steel sheet piling or by the use of ground-water lowering; the injection of cement grouts, silicate solutions or bituminous emulsions can reduce the flow appreciably.

Ground-water lowering methods include shallow-well, well-point, deep-well and multi-stage.

The volume of water to be dealt with in excavations depends on:
(1) the precautions taken to exclude water;
(2) the nature of the ground;
(3) the head causing water to flow into excavations.

Timbering

The Contractor shall supply and fix all necessary timbering, steel sheeting, strutting, shoring, etc., to support the sides of excavations so as to ensure the safety of workmen, freedom from damage of any structures or services and to prevent any movement of adjacent soil. All such supports shall be maintained until the constructional work is sufficiently advanced to permit the timbering, etc., to be withdrawn.

The selection of the method of timbering to be used in excavation work is largely influenced by the type of ground encountered. Four types of sheeting are described in CP 2003: Earthworks, namely poling boards, horizontal sheeting, runners and sheet piling.

Sheet piles are frequently

57

Where necessary, the excavations shall be close timbered or sheeted and the Contractor will receive no additional payment for this work, even although it may have been ordered by the Engineer.

Where the Contractor is required to leave in position timbering or sheeting on the order of the Engineer to safeguard adjoining buildings, etc., he will be paid the schedule price for the materials used. Any timbering or sheeting not ordered by the Engineer but left in for the convenience of the Contractor will not be the subject of any additional payment.

Any bridges, railways, buildings, walls, sewers, culverts, mains, cables, etc., likely to be damaged by the excavation shall be properly supported and the Contractor shall be held responsible for any damage arising in this connection. The Contractor shall be responsible for any damage to the permanent work due to inadequacy of timbering, etc., and any consequential damage caused by the removal of timbering, steel sheeting or other supports from excavations.

interlocking rolled steel sections of various weights and strengths. This type of piling has the great advantage that it can be re-used many times.

Sheet piling is normally driven by a steam- or air-operated hammer. A pile frame or crane is used for pitching the piles and the size of crane or the height of frame required depends on the method to be used for pitching and driving and on the length of pile.

Specialised Work

Tunnel work

Excavation in tunnels and associated shafts shall be performed in a manner approved by the Engineer. All reasonable precautions shall be taken to prevent the subsidence or movement of the surrounding ground or disturbance of adjoining structures.

Where necessary or where directed, the Contractor shall closely timber the

With tunnel work it is imperative that the Contractor takes all necessary steps to prevent any movement of adjacent ground or buildings. Another essential requirement is the provision of adequate timbering. The accompanying specification clauses

excavations to the satisfaction of the
Engineer. When close timbering is not
in use during driving of a tunnel, a set
of timber ready cut and marked for
boxing up the face shall be kept in close
proximity to the work ready for im-
mediate use should an emergency arise.
Where support is required for the ground
at the top and/or sides of a length during excavation, mild steel polings not
less than 10 mm ($\frac{3}{8}$ in.) thick shall be
inserted and left in.

When driving operations cease, the
face shall be closely timbered to the
Engineer's satisfaction and if work
ceases for 48 hours or more, the timbering shall be grouted at the Contractor's
expense.

The Contractor shall include in his
prices for all getting out, filling, wheel-
ing, hoisting, handling, loading, transporting and disposal of the surplus
spoil to the satisfaction of the Engineer.

indicate one way of doing
this.

Use of compressed air plant and equipment.
The Contractor is required to comply
with the regulations contained in the
Report of the Committee on Regula-
tions for the Guidance of Engineers and
Contractors for Work carried out under
Compressed Air, published by the
Institution of Civil Engineers, and of
the Regulations as to Safety, Health and
Welfare in connection with Diving
Operations and Work in Compressed
Air, issued by the Ministry of Labour.

The Contractor shall install, maintain, operate and remove on completion, suitable air compressors, sufficient
to supply adequate compressed air at
the highest pressure that can be

Special precautions are
needed when tunnel work is
to be performed in com-
pressed air. The accompanying specification notes illustrate some of the principal
matters to be considered.
Above all, adequate precautions must be taken to ensure
the safety of the workmen
engaged in this class of work.

required, and allowing sufficient stand-by equipment of not less than one compressor nor less than 20 per cent of the total equipment.

The Contractor shall make arrangements for any alternative standby sources of electrical energy generation or other alternative means to maintain continuity in the event of breakdown of any of the services required for the safety of the workmen, or the Works, or for the prevention of any damage.

The whole of the plant and equipment for the supply of compressed air shall be brought on to the site, erected if possible, and tested before any work in compressed air is commenced.

After erecting and equipping the air locks but before work in compressed air is commenced, the works shall be subjected to a test under a pressure of 0.28 MN/m^2 (40 lbf/in^2) maintained for not less than one hour. The Contractor shall supply, install and maintain in good working order a telephone system connecting the compressor houses, outside and inside of the air locks and tunnel faces, at all times that constructional work is proceeding in compressed air.

Shaft excavation. Under suitable conditions, shaft sinking may be performed by the orthodox underpinning method in free air; excavating, erecting and grouting each ring before the next is commenced. Where necessary, the shafts shall be sunk in compressed air with air locks.

Where steel sheet piling is used, it shall be supported in such a manner

This clause is particularly concerned with the sinking of shaft rings and the use of steel sheet piling to retain the sides of the excavation.

The Contractor may be given the opportunity to use other than orthodox methods subject to certain requirements.

60

that the extreme fibre stress shall not exceed 155 MN/m² (10 tonf/in²). With the consent of the Engineer, sheet piling supports may be in the form of rings of shaft lining, temporarily assembled and suitably packed. Annular space between the outside of the lining and the sheet piling shall be filled with concrete mix C in lifts not exceeding 1·2 m (4 ft) in height and with construction joints in the concrete not less than 150 mm (6 in.) distant vertically from the horizontal (circumferential) joints in the lining.

The Contractor shall include in his prices for all pumping, temporary sumps, etc. needed to control and remove the water entering the shafts.

Where the Contractor wishes to sink a shaft by a method other than that described, he shall submit his proposals to the Engineer, accompanied by adequate drawings and other information, including particulars of the items in the Bill of Quantities requiring adjustment.

Shield-driven work. The main tunnel lengths shall be driven using a hooded Greathead type shield, although tunnels may be driven open-faced for a distance not exceeding 6 m (20 ft) from the shaft break out. This work must be adequately timbered or otherwise supported to the satisfaction of the Engineer, and where directed the timbering or sheeting shall be left in position.

Where no water is encountered, the Contractor shall maintain an air pressure of 0·06 MN/m² or 55 kN/m² (8 lbf/in²) in the tunnel, but where necessary the air pressure shall be increased to balance the hydrostatic head. Where there is insufficient loss of air at

Concrete mix C would probably be 1:3:6.

It is frequently necessary to use a shield in tunnel work, to speed up the job, increase the safety of the workmen and reduce the loss of timbering. Even where shield-driven work is specified, it is customary to permit a short length of tunnel to be driven without a shield when breaking out from a shaft.

the face to give adequate ventilation, the Contractor shall install suitable pipes or ducts for this purpose.

As the shield is driven, auger holes shall be driven in the top of the face for a distance of 600 mm (2 ft) to determine the nature of the ground.

Cofferdams

The Contractor shall provide, construct and maintain proper and sufficient cofferdams of steel sheet piling for dealing with water during the construction of the work adjoining the river bank, and shall remove the cofferdams at completion.

The price for cofferdams shall include for all steel sheet piles, angle and junction piles, struts, cleats, walings, puncheons, tie rods, anchors, etc., and for all labour in driving and fixing and for all clay puddle and other material necessary to make the cofferdams sound and watertight. The method of construction is at the discretion of the Contractor, who is entirely responsible for keeping the cofferdams watertight, but in no case shall the sheet piles be driven to a depth of less than 3 m (10 ft) below the new dredged or excavated levels. The tops of cofferdams shall be not less than 600 mm (2 ft) above flood level.

The Contractor shall be paid billed rates for any steel sheet piling which is left in position on the order of the Engineer.

The cofferdams shall be designed to withstand all pressure conditions obtaining at high and low tides both before and after excavation work has been

The responsibility for the design, provision, maintenance and removal of cofferdams usually rests with the Contractor. They form one of the most important and costly items of temporary works and the accompanying specification clauses list some of the more common requirements.

There is a continual emphasis on the need for watertightness to permit the permanent works within the cofferdam to be constructed in the dry. All responsibilities in connection with cofferdams fall upon the Contractor. The only extra payment he will receive is for sheet piling left in position on the order of the Engineer.

carried out. The cofferdams must be maintained in a stable and satisfactory condition for such periods as are necessary for the performance of the Works, and they are not to be removed without the Engineer's written consent.

In the event of a cofferdam failing, the Contractor is to replace it at his own expense and to the satisfaction of the Engineer. Notwithstanding that the Engineer may have accepted the cofferdam designs submitted by the Contractor, the Contractor will be held entirely responsible for the adequacy and safety of the cofferdams and shall satisfy all claims for damage to new works, damage to property or injury to persons arising out of the failure of the cofferdams and shall indemnify the Employer therefrom.

The Contractor shall keep all cofferdams free from water from any source and shall provide all temporary pumping plant, pumping sumps, subdrains, pipes, channels, etc., and shall fill in any spaces left by the removal of any such works in a manner approved by the Engineer.

No concreting or other constructional work will be permitted unless the excavations are dry and they shall be kept free from water until the concrete has set sufficiently to prevent any possibility of damage by water.

Dredging

The dredging shall be performed to the lines and levels shown on the Contract Drawings or to such other lines and levels as may be directed by the Engineer.

Due to the cumbersome machinery used and the fact that the excavation is carried out below water level, it is customary for tolerances

The work shall be undertaken with a suitable dredger and the excavated material shall be removed to a site approved by the River Authority.

Dredging shall be paid for on the basis of measurement in barge. The Contractor will be permitted a tolerance of 300 mm (12 in.) on horizontal surfaces and 600 mm (24 in.) on sloping surfaces, below the specified dredged levels. The volume of any excess dredging over this tolerance will be deducted from the barge quantities.

The Engineer and the Contractor shall undertake and agree a joint post-dredging survey to determine the final dredged levels.

below specified dredged levels to be allowed. Measurement of dredged quantities is usually by barge or hopper, but a check on final dredged levels by soundings will need to be made.

CHAPTER FIVE

Specification of Concrete Work

ALL civil engineering jobs make use of concrete in one way or another. This chapter therefore covers an important section of civil engineering work. The drafting of specification clauses for concrete piles and roads is dealt with in later chapters.

It is advisable to adopt a logical sequence of items in this work section. These can be conveniently grouped under six main headings, namely: materials; concrete work; reinforcement; shuttering; precast work; and prestressed work. In practice the actual order adopted varies from office to office, but some advantages would accrue from the general adoption of a standardised order of items. Within each subsection, the items should also follow a logical order and, in the case of concrete, these can conveniently follow the order of operations on the site.

A selection of typical clause headings, grouped under the four main sections of reinforced concrete work, are now listed.

MATERIALS

(1) Cement
(2) Fine aggregate
(3) Coarse aggregate
(4) Water
(5) Reinforcement
(6) Samples and tests
(7) Stocks of materials

Concrete Work

(1) Concrete mixes
(2) Weigh-batching
(3) Mixing concrete
(4) Workability of concrete
(5) Test cubes
(6) Percolation tests
(7) Transporting concrete
(8) Placing concrete
(9) Blinding coat
(10) Vibrators
(11) Construction joints
(12) Expansion joints
(13) Joining new and old concrete
(14) Surface finish to concrete
(15) Concreting in cold weather
(16) Curing concrete
(17) Fixing bolts, etc.
(18) Pipes through concrete walls
(19) Tests for watertightness
(20) Concreting records

Reinforcement

(1) Bending reinforcement
(2) Placing reinforcement
(3) Cover to reinforcement

Shuttering

(1) Shuttering generally
(2) Shuttering to beams and slabs
(3) Preparation of shuttering
(4) Striking shuttering

Precast Concrete and Prestressing Work

Typical specification clauses under these headings follow, although it is emphasised that the actual requirements vary considerably from one job to another.

MATERIALS

Cement

Unless otherwise specified or ordered by the Engineer, the cement shall be ordinary Portland cement complying with B.S.12. The cement shall be delivered either in unbroken bags of the manufacturer and stored in a waterproof shed with a raised boarded floor, or delivered in bulk for bulk storage, provided that the Engineer is satisfied that the methods of transport, handling and storage are satisfactory.

In both cases the cement shall be stored in such a way that each consignment shall be used in order of receipt. Each consignment of cement shall be delivered to the site at least two weeks before it is required for use and the Contractor shall supply the Engineer with a copy of the manufacturer's test certificate for each consignment.

The special conditions relating to the storage and use of rapid-hardening cement shall be strictly observed and different types of cement shall be kept separate at all times.

It is essential that all cement used complies with the appropriate British standard. It is customary to permit delivery in bags and storage in suitable sheds, or delivery in bulk in specially designed vehicles and storage in suitable bins or silos.

On delivery fresh cement may be at a high temperature and for this reason it is customary to require cement to be stored on the job for at least two weeks prior to use. Conversely, the Contractor will not be permitted to use stale or lumpy cement.

Fine aggregate

Fine aggregate shall be well washed and shall be sharp and free from clay, chalk, organic matter and other impurities. It shall comply with the requirements of B.S. 882 and shall be graded in accordance with Table 2, Zone 2, namely:

The most usual procedure is to require compliance with B.S. 882 and such additional requirements as the Engineer thinks fit.

67

To pass sieve size	Percentage
10·00 mm	100
5·00 mm	90–100
2·36 mm	75–100
1·18 mm	55– 90
600 μm	35– 59
300 μm	8– 30
150 μm	0– 10

Coarse aggregate

Coarse aggregate shall be gravel or other suitable material and shall be well washed and free from sand, clay, quarry refuse and other impurities. It shall comply with the requirements of B.S. 882 and shall be graded in accordance with Table I, namely:

Similar provisions inserted as for fine aggregate. These percentages apply to aggregates of 20 to 5 mm in size.

To pass sieve size	Percentage
37·5 mm	100
20·0 mm	95–100
10·0 mm	30– 60
5·0 mm	0– 10

Water

Water supplied by the local water undertaking only shall be used for mixing concrete, mortar and grout. It shall be free from organic or other harmful substances in solution or suspension, and shall be tested for suitability in accordance with B.S. 3148.

It is advisable to restrict the water used on the job to that supplied by the local water undertaking to ensure a reasonable standard of purity.

Steel reinforcement

Bar reinforcement shall be mild steel round bars of British manufacture by the open hearth (acid or basic) process

Specification clauses for reinforcement invariably refer the Contractor to British

and shall comply with the requirements of B.S. 785. The Contractor shall supply the Engineer with mill order sheets and test certificates.

Fabric reinforcement shall conform to B.S. 4483 and shall be supplied in flat sheets.

All reinforcement shall be free from oil, grease, dirt, paint and any loose rust prior to use.

Standards for detailed requirements. This procedure shortens the specification (which is always a lengthy document) and yet, at the same time, ensures a high standard of material, well suited for the job in hand.

Samples of aggregate

Samples of the aggregate which the Contractor proposes to use on the Works shall be deposited with and approved by the Engineer prior to commencement of the Works. All aggregates used shall be equal to the original samples and further samples shall be supplied as required.

Some Engineers require the Contractor to carry out grading analyses, voids tests, bulk density tests, silt tests and tests for organic impurities, and to supply the necessary equipment for this purpose.

Stocks of cement and aggregates

The Contractor shall maintain on site stocks of cement and aggregate to cover not less than two weeks' requirements.

This clause avoids the possibility of the job being retarded owing to lack of essential materials.

CONCRETE WORK

Concrete proportions

Concrete for reinforced work shall comply with the requirements of British Standard Code of Practice CP 114: Structural Use of Reinforced Concrete in Buildings, unless this specification contains different requirements. The following table indicates the mix requirements for reinforced work.

B.S. Code of Practice 114 provides a good basis for the construction of reinforced concrete work. Mixes can be specified by weight, volume, strength or a combination of these.

Nominal mix	Class	Minimum cube strength: N/mm^2 ($lbf/in.^2$))	
		Preliminary tests at 28 days	Works tests at 28 days
1:1:2	A	40	30
1:1½:3	B	34	25·5
1:2:4	C	28	21

The proportions of fine to coarse aggregate and of cement to combined aggregate will depend on the type and grading of the aggregate and shall be determined in accordance with the recommendations contained in Road Research Paper No. 4, to secure concrete of the required strength and of the highest possible density.

Mass concrete shall consist of cement and 'all-in' aggregate of the following mixes:

This table lists the nominal mixes of concrete with their class references and gives the minimum permissible crushing strengths for each mix.

Exact proportions of aggregates are dependent on a number of factors and will have to be decided on the job.

'All-in' aggregate may consist of river ballast or a mixture of sand and coarse aggregate graded in accordance with Table 3 of B.S. 882, with a nominal maximum size of 40 mm (1½ in.).

Nominal mix	Class	Cement: kg (lb)	'All-in' aggregate (dry): m^3 (ft^3)	Minimum cube strength: N/mm^2 ($lbf/in.^2$)	
				Preliminary tests at 28 days	Works tests at 28 days
1:6	D	50 (112)	0·21 (7½)	21 (3000)	17 (2500)
1:8	E	50 (112)	0·28 (10)	16 (2300)	12 (1700)

Gauging of concreting materials

Fine and coarse aggregates for structural work shall be measured in a weigh-

The bulk of concrete ingredients are measured by

batching machine of approved design. The machine shall be maintained in good working order with periodic checks for accuracy.

The cement shall be gauged by weight, using one or more 50 kg (112 lb) bags of cement to each batch. The water for each batch shall be measured by volume in a calibrated container.

The Engineer may permit the use of gauge boxes for the measurement of aggregates for non-structural concrete, such as in blinding coats and surrounds to pipes. Due allowance must be made for the bulking of the fine aggregate and the boxes shall be capable of use without dividing the contents of bags of cement.

weigh-batching plant, which is much more convenient, less laborious and makes for greater accuracy in use, as compared with gauge boxes. Nevertheless, it may be permissible to allow the use of gauge boxes for small quantities of concrete or where the standards are less exacting.

Mixing of concrete

Concrete shall be mixed in batch type mixers of approved design. The volume of materials inserted per batch shall not exceed the manufacturer's rated capacity and the volume of each batch shall be such that only whole bags of cement will be used. The mixer drum shall be emptied completely before being refilled. All materials shall be mixed until the concrete is uniform in colour and consistency and in no case shall it take less than two minutes.

At commencement and on completion of each mixing period, the drum of the mixer shall be thoroughly washed out with clean water and it shall be kept free from hardened or partially set concrete.

Under special circumstances the Engineer may permit hand-mixing of concrete. The ingredients shall be mixed

The majority of specifications require concrete to be mixed in approved batch type mixers. The accompanying specification clauses list the principal precautions to be taken if good quality concrete is to be produced.

Where small quantities of non-structural concrete are required the Engineer may permit the use of hand-mixed concrete, subject to certain conditions.

71

dry on a watertight platform until a uniform colour is obtained. Clean water shall then be added gradually through a rose-head and the whole mass turned over at least three times in a wet state until it attains a slightly wet consistency. A 10 per cent reduction shall be made in the quantities of fine and coarse aggregate at the Contractor's expense and the concrete shall be carefully checked by slump test or compacting factor test.

Water content and consistency

Clean water shall be added in the quantity required to maintain the water/cement ratio at the optimum value as determined from the preliminary tests, to secure a sufficiently impervious concrete of adequate strength and workability, in accordance with CP 2007.

The Contractor shall check the moisture content of the aggregates in determining the volume of water to be added to each batch of concrete. The Contractor shall keep sufficient equipment on the site for carrying out slump tests and/or compacting factor tests during each day of concreting in the manner described in B.S. 1881 and shall keep a record of these tests.

The following table gives a guide as to probable test limits.

The water content has to be kept within narrow limits to ensure a dense concrete of adequate strength and impermeability. B.S. 1881: Methods of Testing Concrete indicates two ways of testing the concrete for correct consistency: slump test and compacting factor test.

The accompanying specification clause also indicates typical limits for slumps and compacting factors for different classes of work. The compacting factors can also vary with different mixes.

Hand compacted concrete:	Slump	Compacting factor
Mass concrete filling and blinding	40–65 mm ($1\frac{1}{2}$–$2\frac{1}{2}$ in.)	0·89–0·93
Reinforced concrete foundations, floors, beams and slabs	50–75 mm (2–3 in.)	0·91–0·94
Reinforced concrete columns and walls	65–90 mm ($2\frac{1}{2}$–$3\frac{1}{2}$ in.)	0·93–0·95

Vibrated concrete:	Slump	Compacting factor
Reinforced concrete floors, beams and slabs	12–25 mm ($\frac{1}{2}$–1 in.)	0·82–0·85
Reinforced concrete columns and walls	25–50 mm (1–2 in.)	0·85–0·91

Test cubes

The strength of concrete shall be determined by tests on cubes made, cured and tested in accordance with B.S. 1881 (Parts 7 and 8), except that the temperature during the first two weeks of curing shall be between 15 and 21°C (60 and 70°F).

Six cubes shall be taken for each section of the work during each half-day's concreting. If the minimum batch cube strength (average strength of batch less twice the value of the standard deviation of the results) is less than the specified minimum strengths, the concrete represented by these cubes shall be cut out and replaced with satisfactory concrete at the Contractor's expense.

Provision must be made for the taking of a sufficient number of cubes of concrete to maintain a check on the crushing strength of the concrete. The method of applying the results of the tests must also be indicated.

Percolation tests

Concrete in structures which are to withstand water under pressure is to be the subject of percolation tests. The Contractor shall supply concrete test slabs 125 mm (5 in.) in diameter and 50 mm (2 in.) thick, gauged in the specified proportions, to a testing laboratory, where they will be subjected to a

There are occasions when concrete offering a high degree of resistance to water penetration is necessary, as with circulating water ducts on power stations. In these circumstances it is desirable to make provision for test water pressure equivalent to 12 m (40 ft) head of water on one side of the slab for 24 hours. Should dampness appear on the other side of the slab, further test slabs shall be prepared with adjusted proportions of mix. This procedure shall be repeated until slabs are produced which satisfactorily meet the requirements of the test and the mix adopted in the last test will be used throughout this class of work.

slabs of concrete to be prepared and subjected to percolation tests, with a view to determining a suitable mix of concrete for this class of work.

Transporting concrete

All concrete shall be transported from the mixer to the place of final deposit as speedily as possible, and in no case shall this exceed 20 minutes after mixing. The method of transit shall be such that it will prevent the segregation, loss or contamination of the ingredients.

The concrete must be transported to its final position by a satisfactory method as speedily as possible.

Placing concrete

Before any concrete is placed in position the shuttering and other adjoining surfaces shall be clean and free from all foreign matter. Care must be taken to prevent workmen placing concrete from introducing clay or other harmful matter on their boots.

The concrete shall be thoroughly worked into all parts of the shuttering and between and around the steel reinforcement, and compacted by approved methods to give a dense and compact concrete, free from voids of any kind. Great care shall be taken to prevent the displacement or deformation of the steel reinforcement during concreting.

This forms an important part of any specification dealing with concrete work, as special care must be taken in placing and compacting concrete if the best results are to be achieved. Essential precautions, such as keeping surfaces of shuttering clean, preventing the displacement of steel reinforcement, depositing the concrete in relatively thin horizontal layers and finishing at construction joints all need emphasis.

Concrete placed against shuttering to form an exposed surface shall be particularly well vibrated or otherwise compacted to produce a perfectly smooth finish.

Concrete shall be deposited in layers not exceeding 225 mm (9 in.) in thickness and the surface of all concrete during depositing shall be kept reasonably level. No concrete shall be allowed to fall uncontrolled through a height of more than 1·25 m (4 ft).

No concrete shall be deposited in water except where indicated on the drawings or where special permission is granted by the Engineer. Where permissible, the concrete shall be placed by tremie pipe or lowered in boxes with opening bottoms, in bags or by other approved method.

Concreting shall be carried out continuously between construction joints with each section completed in a single working day unless specially authorised by the Engineer.

Lifts of concrete shall normally be not less that 600 mm (2 ft) nor greater than 2 m (6 ft) in height. In the event of unavoidable stoppages at positions other than those required, the concrete shall be terminated on horizontal planes and against vertical surfaces, and construction joints shall be formed in these positions. After being placed in position the concrete shall not be subjected to any disturbance other than that associated with compacting.

Where screeds, rendering or granolithic finish are to be applied, the surface of the concrete shall be left rough to form a key. Where a smooth floated finish is required, care shall be taken to

Detailed requirements will vary from job to job, but the clauses given form a useful guide and incorporate the more usual requirements.

avoid an excess of water in the top layer of concrete. Where excess moisture arises, a dry mixture of cement and fine aggregate in the proportions used in the concrete shall be sprinkled on the surface and worked in with a float.

Blinding coat

Reinforced concrete shall not be laid directly onto earth surfaces. A blinding coat of 75 mm (3 in.) minimum thickness of concrete class D shall be laid on the ground before any reinforcement is placed in position.

It is usual to specify a thin blinding coat of mass concrete to provide a working platform for the reinforced concrete work which follows.

Vibrated concrete

All reinforced concrete of classes A and B shall be compacted with approved insertion vibrators operating at a frequency of not less than 5000 c/min. The vibrators shall be operated by men skilled and experienced in this class of work.

Care shall be taken to prevent contact between the vibrators and the reinforcement, and to ensure that the concrete is not vibrated in a manner likely to cause damage to previously placed concrete. Vibration shall be discontinued as soon as water or grout appears on the top of the concrete.

It is a common specification requirement that reinforced concrete in structural members shall be vibrated to ensure maximum compaction. It is necessary to insert clauses to ensure the use of the right type of vibrator in a manner which will not produce adverse effects. Vibrators of the needle or shutter types are often specified.

Construction joints

Construction joints shall be formed only in positions approved by the Engineer and the Contractor shall accept full responsibility for the soundness of such joints.

It is important that the location of construction joints is approved by the Engineer, so that these only occur in places where they

76

Vertical construction joints in walls and construction joints in slabs shall be formed with stop shutters, holed or slotted for reinforcement. Where joints are visible on the finished face, a 25 mm (1 in.) square lath shall be attached to the main shuttering in order to leave a clean, straight joint on the face of the finished work. Joggles 50 mm (2 in.) deep shall be formed at construction joints in walls to water-retaining structures of 225 mm (9 in.) or greater thickness.

Walls and floors exceeding 10·5 m (35 ft) in length shall be constructed in alternate bays with each bay being not more than 10·5 m (35 ft) square. Intermediate bays shall not be placed until at least 28 days after the adjoining concrete has been laid.

As soon as the concrete has attained its final set, the surface of the construction joint shall be wire brushed or hacked as appropriate to remove all laitance and expose the aggregate. A coat of cement mortar (1:2) 15 mm ($\frac{1}{2}$ in.) thick shall then be applied to the washed and cleaned surface and concreting shall follow within 20 minutes.

cause the least possible harm. For instance, in the case of reinforced continuous slabs, the construction joints should be at the centre of a support over which such members will be continuous. In the case of beams, the joints should be either over a support or at the centre of the span.

The accompanying clauses also outline typical requirements for forming the joints and for treating the surface of the joint prior to placing further concrete.

Expansion joints

Expansion joints shall be formed of approved non-extruding jointing material sealed with a hot-poured bituminous sealer.

Proprietary materials are often specified for this work.

Surface finish to concrete

The Engineer's representative shall inspect all concrete faces after shuttering

Provision is made for rectifying the smaller defects

has been struck and before any work is done to the surfaces of the concrete.

All surfaces to be permanently exposed shall be rubbed down to remove all fins and other projections and all angles shall be finished satisfactorily. The rubbing down shall be performed with power-driven grinding wheels where necessary.

Any honeycombed surfaces accepted by the Engineer shall be filled with cement mortar having the same proportions of cement and fine aggregate as the concrete and shall be finished to a true surface with a float.

occurring in concrete surfaces after the striking of shuttering. On occasion, the Contractor is also required to wash all exposed surfaces with water and then to coat them with a cement wash (1 part cement: $1\frac{1}{2}$ parts fine sand) rubbed in with carborundum blocks and steel floats.

Concreting in cold weather

The Contractor shall provide sufficient thermometers for measuring the temperature of the air, mixing water, aggregates, finished concrete, etc. No concreting shall be carried out when the air temperature is below 4°C (40°F) or 1°C (34°F) up to 2 hours before sunset subject to the following precautions being taken:

(1) Cement shall be properly stored.
(2) Aggregates shall be protected from frost by tarpaulins and heated if necessary.
(3) Mixing water shall be free from ice and heated if necessary.
(4) Any frost, snow or ice shall be removed from shuttering.
(5) No delay shall occur between mixing and placing concrete and no concrete shall be left unplaced in break periods.
(6) All exposed surfaces of concrete, metal shuttering and projecting reinforcement shall be adequately

In the past, the time lost on civil engineering jobs during very cold weather was considerable.

In recent years an endeavour has been made to find methods of permitting concreting and other work to proceed for longer periods during the winter months.

The provisions described aim at ensuring that the temperature of the concrete is not less than 10°C (50°F) at the time of placing and is kept above 4°C (40°F) for three days afterwards, or not less than 4°C (40°F) on placing and above 2°C (35°F) for seven days afterwards. The latter periods can be reduced if rapid-hardening cement is used.

covered as soon as practicable, with a 150 mm (6 in.) air space left between the concrete and the covering.

(7) Curing times shall be extended by the number of days on which the temperature falls below 2°C (36°F) unless the Engineer permits the addition of 1 kg (2¼ lb) of calcium chloride per 50 kg (cwt) of cement.

Curing concrete

All exposed concrete surfaces shall be covered with hessian or other suitable material, and this shall be kept moist for at least 14 days after the placing of the concrete.

Curing is essential to prevent the formation of hair cracks on the surface of the concrete caused by the speedy drying action of sun and winds.

Fixing bolts, etc.

Anchor bolts, Lewis bolts, ragbolts, tubes, steel angles, etc., shall be built into the concrete as shown on the Drawings or as directed by the Engineer. Where necessary, temporary wooden boxes shall be set in the concrete and so arranged that the box may be withdrawn and the bolt or other article set in the recess so formed and grouted up solid. The Contractor is to include in his price for temporary boxings and for keeping the recesses free from oil, grease and debris.

Bolts are frequently required to be set in the concrete as fixings for machinery, pipes, etc. Two alternative procedures are available: building the bolts into the concrete as it is placed, or fixing temporary boxings to produce holes into which the bolts can be grouted subsequently.

Pipes through concrete walls, etc.

Pipes shall be built through concrete walls and slabs wherever possible at the time the concrete is placed. Boxing-out for subsequent insertion shall only be

The main requirements are that the pipes should be accurately positioned and so fixed as to secure watertight

79

adopted with the approval of the Engineer, and the Contractor shall be entirely responsible for securing sound and watertight construction.

The Contractor shall take the greatest care to set the pipes in the correct positions and to the required inclinations. In this connection the Contractor shall collaborate to the fullest possible extent with any plant contractors involved.

Tests for watertightness of tanks

After at least 28 days from completion of construction of concrete water-retaining structures, but before backfilling earth around them, the structures shall be filled with water. After sufficient time has been allowed for absorption, the water level shall be measured daily for a seven-day period, extended if necessary for periods of rain.

If leakage occurs, the Contractor shall be required to make good any defects and repeat the test for watertightness at his own expense. The Contractor must allow in his price for the supply and removal of the water, sealing and unsealing of pipes, and all other expenses to be incurred in carrying out the tests.

Concreting records

The Contractor shall be required to keep the following records.
(1) Daily minimum and maximum temperatures.
(2) Weather conditions when concreting is proceeding with reference to moisture, winds and sunshine.

joints. Some Engineers do not favour boxing-out for pipes but this procedure is unavoidable on occasions.

With reservoirs and tanks for the storage of sewage, water, etc., it is essential that the structures should be watertight, and tests are necessary to establish that this is so.

These records may subsequently prove valuable if parts of the concrete prove to be sub-standard, as they may provide a guide to the probable cause.

(3) Part of structure concreted each day, shown on a set of drawings, and amount of water added to each batch of concrete suitably recorded.

REINFORCEMENT

Bending reinforcement

All bar reinforcement shall be bent cold, before the bars are placed in position. No heating or welding will be permitted.

Bar reinforcement shall be shaped to the exact dimensions required. All bending dimensions and tolerances shall comply with B.S. 4466. Hooked ends shall each have an internal diameter of curvature and a straight length beyond the semi-circle of at least four times the diameter of the bar.

It is a usual requirement that all bar reinforcement shall be bent cold. It is also useful to refer to B.S. 4466: Bending Dimensions and Scheduling of Bars for the Reinforcement of Concrete, for bending dimensions and tolerances.

Placing reinforcement

All reinforcement shall be set out exactly as shown on the working drawings. It shall be supported by an adequate number of small precast concrete blocks with tying wire cast in, or with other approved spacers. The bars shall be adequately tied at intersections with 18 gauge annealed steel wire.

All joints in bar reinforcement shall overlap for a distance of at least 40 times the diameter of the smallest lapped bar. With fabric reinforcement the following laps shall operate:
oblong mesh: 450 mm (18 in.) along
                 longitudinal wires
               75 mm (3 in.) along
               transverse wires

The main points to cover are the setting out of the bars, the method of fixing and the amount of laps. In particular, pieces of steel or blocks of wood should not be used to keep the reinforcing steel permanently in position.

81

square mesh: 300 mm (12 in.) in both
directions

Cover to reinforcement

Reinforcing bars shall be provided with the following minimum cover of concrete:

It is essential that steel reinforcement in structural members should be given sufficient concrete cover to ensure adequate strength and prevent the rusting of the reinforcement. The minimum amount of cover required varies with the type of member, size of bar and position of the work.

(1) To each end of a reinforcing bar, not less than 25 mm (1 in.) nor less than twice the diameter of the bar.

(2) To a longitudinal reinforcing bar in a column, not less than 40 mm (1½ in.) nor less than the diameter of the bar.

(3) To a longitudinal reinforcing bar in a beam, not less than 25 mm (1 in.) nor less than the diameter of the bar.

(4) To a reinforcing bar in a wall, not less than 20 mm (¾ in.) nor less than the diameter of the bar.

(5) To a reinforcing bar in a slab, not less than 15 mm (½ in.) nor less than the diameter of the bar.

(6) To any other reinforcement, not less than 15 mm (½ in.) nor less than the diameter of the bar.

All the above dimensions shall be increased by 15 mm (½ in.) for external faces of concrete exposed to the weather and for all faces in contact with earth. Where external faces of concrete are exposed to water, the cover of concrete shall be not less than 50 mm (2 in.).

SHUTTERING

Design and construction of shuttering

All shuttering, whether of timber or metal, shall be in every respect adapted to the structure and to the required

The responsibility for providing adequate shuttering rests entirely with the

surface finish of the concrete. All shuttering shall be fixed in perfect alignment and be securely braced to withstand, without appreciable displacement, deflection or movement of any kind, the weight of the construction and the movement of persons, materials and plant. Notwithstanding approval by the Engineer, the strength and adequacy of the shuttering shall remain the responsibility of the Contractor. All joints shall be sufficiently close to prevent leakage of liquid from the concrete. Wedges and clamps shall be used wherever practicable in the construction of the shuttering to permit easy adjustment and removal. Where special surface treatment is required, the shuttering shall be lined with hardboard.

All pipes, angles, etc., that are required to be built in or through the concrete shall be firmly fixed in the shuttering, which shall be neatly and accurately cut and fitted around them. The joints shall be caulked where necessary to prevent leakage of grout or fines.

Contractor. Nevertheless, it is advisable for the Engineer to give guidance to the Contractor as to the precautions to be taken, thus possibly avoiding failure of the shuttering and consequent delays to the work.

Where a perfectly smooth finish is required to the concrete, as to the interior surfaces of circulating water ducts, it is customary to specify the use of metal faced moulds and sheeting.

Shuttering to vibrated concrete

Where concrete is to be vibrated and timber shuttering is used, all joints shall be tongued and grooved or the boards shall have straight parallel edges planed perpendicular to the board surface. Where necessary, to prevent leakage of grout or fines, joints shall be caulked with putty or other approved material before concreting is commenced.

Where steel shuttering is used, all joints and holes in the shuttering shall be sealed with tape held in position with spirit or other suitable glue.

Shuttering to vibrated concrete has to withstand the more severe effects resulting from the vibration. Special attention has to be directed to the joints in the shuttering.

Shuttering to beams and slabs

Beam shuttering shall be designed so that the sides may be removed without disturbing the bottom boards or their supports. The bottom boards shall be set to a camber approximating to the final deflection as determined by the Engineer. The supporting struts shall be adjusted in position and be suitably supported at their lower ends on proper soleplates.

Boarded shuttering to the soffit of slabs shall be laid perfectly true and adequate bearers and struts shall be provided.

It is sometimes specified that the bottoms of beam shutters should have a camber of from 1/300 to 1/350 of the span according to the size of beam. Beam shutters should be so constructed that the only parts not immediately removable are those directly above the supporting struts.

Preparation of shuttering

Where narrow members of considerable depth are to be concreted, the Contractor shall, if directed by the Engineer, provide temporary openings in the sides of the shuttering to facilitate the pouring and compacting of the concrete. Small temporary openings shall be provided as necessary at the bottoms of shuttering to columns, walls and deep beams to permit the extraction of debris.

Before concreting is commenced, shuttering and centering shall be carefully examined and cleaned out. The inside surfaces of the shuttering shall be coated with approved mould oil to prevent adhesion of the concrete.

It is essential that all shutters should be examined and cleaned out immediately before concreting is started. Adequate provision must be made for access points for the removal of debris. In addition to coating the interior surfaces of shuttering with mould oil, it is a common practice also to require their wetting to prevent absorption of the water from the concrete.

Striking shuttering

No shuttering or supports for shuttering shall be struck or moved without the Engineer's consent. The work of

The sole responsibility for the removal or striking of shuttering and supports, and removing the shuttering or supports shall be carried out under the personal supervision of a competent foreman.

The Contractor shall be responsible for any injury to the work and any consequential damage caused by or arising from the moving or striking of shutters or supports. Any advice, permission or approval given by the Engineer's representative shall not relieve the Contractor of his responsibilities.

All shuttering shall be removed without shock or vibration and in the manner and order approved by the Engineer. Before the shuttering is stripped, the concrete surface shall be exposed where necessary in order to determine whether the concrete has hardened sufficiently.

In general, the following minimum times shall elapse between concreting and striking of shuttering:

in particular its premature removal, rests with the Contractor.

Some Engineers give guidance on possible minimum striking periods, such as those given by the author. These periods will be reduced if rapid-hardening cement is used.

Striking periods are sometimes calculated on the following basis for ordinary Portland cement concrete and normal weather conditions:
Bridge abutments and wing walls: 5 days
Archwork: 28 days
Roof, slabs, beam soffits, etc.: 7 days plus $\frac{1}{3}$ day (1 day) for every metre (foot) of span over 2 metres (6 feet).
Beam sides: 3 days.

	Cold weather (about freezing point) Days	Normal weather (about 16°C (60°F) Days
Beam sides, walls and columns (unloaded)	8	2
Slabs for X m (ft) span	$\frac{1}{2} X$ ($1\frac{1}{2} X$) but not less than 5 (15)	X but not less than 3·5 (10)
Beam soffits for Y m (ft) span	$\frac{1}{2} Y$ ($1\frac{1}{2} Y$) but not less than 10 (30)	Y but not less than 7 (20)

PRECAST CONCRETE

All precast concrete members shall be cast on the site on a suitable casting platform and in accordance with the

Alternatively, the Engineer may be prepared to allow the casting of precast members

detailed drawings. Steel reinforcing bars shall be left projecting from the ends of precast members for building into in situ concrete work.

The Contractor shall allow in his price for all necessary moulds, handling, transporting, hoisting and lowering the precast members into position, temporarily supporting as necessary and fixing to the satisfaction of the Engineer.

off the site, although he will require access to the casting area.

PRESTRESSED WORK

Prestressing equipment and materials

All prestressing steel shall be free from loose rust, loose mill scale, oil, grease or any other harmful material. Cables shall be free from twists and shall have tags affixed to them indicating the cable and coil or steel numbers.

All sheaths and cores shall be placed and maintained in their correct positions while the concrete is being placed. Cores shall not be extracted until the concrete has set sufficiently hard to permit withdrawal without damage. Joints in sheaths shall be kept to a minimum and shall be properly sealed. Ducts through adjoining stressed units shall be in perfect alignment with one another and with the anchorage assemblies throughout.

Several proprietary systems of prestressing are in operation. This makes for difficulties in specifying the materials and method of stressing in general terms. Nevertheless, there are some general rules applicable to all systems and these have been brought out very effectively in the Specification for Road and Bridge Works issued by the Ministry of Transport. This document has been used as a guide in framing the accompanying specification clauses.

Tensioning procedure

Tensioning shall be carried out only in the presence of and to the approval of the Engineer. No member shall be stressed until the concrete is of the required age as shown by test cubes.

The tendons shall be stressed gradually and evenly until they attain the required stress and shall then be anchored. After the tendons have been anchored, the jack pressure shall be released gradually and evenly. The Contractor shall keep records of all tensioning operations, including measured extensions, pressure-gauge and load-meter readings and amount of pull-in at anchorages.

Grouting of tendons cannot proceed until ducts have been cleaned and anchorages sealed with mortar. The grout injection equipment shall be able to operate continuously, recirculate grout and attain a delivery pressure up to 0.7 MN/m^2 (100 lbf/in.^{2}). The grout shall consist of Portland cement and water with a water/cement ratio not exceeding 0.45.

Where members are tensioned off the site, the Contractor shall notify the Engineer of the starting dates of the various operations and shall supply weekly reports giving details of units cast and stressed and of test results.

CHAPTER SIX

Specification of Brickwork, Masonry and Waterproofing

THIS chapter deals with the drafting of specification clauses covering brickwork, masonry and waterproofing work. It is desirable that a logical sequence of items should be adopted for each section of work, preferably beginning with materials clauses and following with particulars of workmanship. It will be appreciated that the detailed requirements vary from job to job and that the typical clauses produced aim at giving a guide as to their possible form and content.

There are a number of relevant British Standards and Codes of Practice which can be usefully employed and reference is made to most of them in the typical specification clauses which follow. The principal items to be covered in specification clauses for each of the three works sections covered by this chapter are now listed.

BRICKWORK

(1) Bricks
(2) Mortars
(3) Bricklaying generally
(4) Damp-proof courses
(5) Faced brickwork and pointing
(6) Special features or classes of work

MASONRY

(1) Dressed stonework (ashlar)
(2) Various forms of rubble walling (squared, random, dry rubble walling, etc.)

(3) Special stonework (copings, quoins, arch stones, etc.)
(4) Dowels, cramps, etc.
(5) Cast stonework

WATERPROOFING

The nature and content of the specification clauses in this section will be largely determined by the constructional details of the particular job. Asphalt is widely used in good-class work and various bituminous applications are also available, in addition to polythene sheeting, there is a wide range of proprietary integral waterproofers and waterproofing paints. In each case, the specification clauses can normally be sub-divided into two categories: materials and workmanship. The workmanship clauses will often describe the method of performing the work as well as the quality required.

TYPICAL SPECIFICATION CLAUSES   EXPLANATORY NOTES

BRICKWORK

Bricks generally

All bricks shall conform to the requirements of B.S. 3921 and in addition they shall be hard, sound, square, well burnt, uniform in texture, regular in shape, with true square arrises, and even in size. Care is to be taken in unloading, stacking and handling and no chipped or damaged bricks shall be used.

All bricks shall be equal to samples submitted to and approved by the Engineer before any brickwork is commenced. Adequate stocks of bricks shall be maintained on the site to ensure continuity of working.

Common bricks. The common bricks shall be 65 mm (2⅝ in.) commons supplied by Messrs X or other equal and approved.

All bricks, irrespective of their type or function, must conform to certain basic minimum requirements and be equal in all respects to submitted and approved samples.

The common bricks are often obtained from a local brickworks when one is available.

Engineering bricks. Engineering bricks shall conform to the requirements for class B engineering bricks in B.S. 3921.

This British Standard details two classes of engineering brick with minimum average compressive strengths of 69·0 N/mm² (10,000 lbf/in.²) for class A bricks and 48·5 N/mm² (7000 lbf/in.²) for class B bricks.

The standard brick size is 215 × 102·5 × 65 mm and with 10 mm joints gives a unit size of 225 × 112·5 × 75 mm.

Facing bricks. Facing bricks are to be Antique dark brown facing bricks obtained from Messrs *Y* at a prime cost of £45·0 per thousand bricks delivered to the site.

In most cases the facing bricks will have been selected to ensure that the desired colour and texture of brick is obtained. The inclusion of a prime cost prevents each of the contractors tendering having to obtain quotations.

Mortar

Mortar for engineering brickwork shall consist of one part of ordinary Portland cement, as previously specified, to three parts of sand. The sand shall comply with B.S. 1200, Table 1, shall be approved by the Engineer before use and shall be adequately protected from contamination.

Sand shall withstand satisfactorily the following test. A sample of sand shall be mixed with water (in the proportions of one part sand to two parts water) for one minute in a cylindrical glass jar with a flat base. It shall then be set aside for two hours for the contents to settle and at the end of this period, the top layer of fine materials shall not exceed 5 per cent of the volume of the solid matter in

Different types and proportions of mortars may be specified for different classes of work. Strong cement mortars are needed for heavy load-bearing and damp-resistant structures, gauged mortar for general work and possibly lime mortar for work requiring maximum flexibility, as in tall chimney shafts.

All the materials used should comply with the relevant British Standards and adequate safeguards should be introduced to cover gauging, mixing, sampling and use of mortar.

the jar. The water remaining from the test shall show no evidence of sewage, organic or chemical contamination.

Different sands shall be stored separately and sands for pointing shall be obtained in sufficient quantity at one time to enable material of the approved colour to be used for the whole of the work. Sand for pointing mortar and fine joint work shall pass a 1·5 mm ($\frac{1}{16}$ in.) square mesh sieve.

Mortar for common and faced brickwork shall consist of gauged mortar mixed in the proportions of one part of ordinary Portland cement, one part of non-hydraulic or semi-hydraulic lime and six parts of sand. The lime shall comply with B.S. 890, class B, and shall be stored in a similar manner to that described for cement.

All materials for mortar are to be accurately measured in proper gauge boxes and shall be mixed on a suitable stone or wooden banker or in an approved mixer with only sufficient water added to produce a workable mix. All gauge boxes, bankers and mixers shall be kept clean.

The Contractor shall permit samples of mortar to be taken from time to time for testing and re-tempering of mortar will not be permitted. All mortar shall be mixed in quantities sufficient for only 30 minutes work.

Bricklaying

All brickwork shall be built to the dimensions, thicknesses and heights and in the positions shown on the Drawings or as directed by the Engineer and in conformity with CP 121 Part I. Clean off

It is usual to specify that all brickwork shall be laid to the dimensions and in the positions shown on the Drawings. It is also advisable to

and prepare all concrete and other surfaces on which bricks are to be laid.

. All brickwork shall be built uniform, true and level, with all perpends vertical and in line, and shall rise 300 mm (12 in.) in every four courses. No brickwork shall rise more than 1·25 m (4 ft) above adjoining work during bricklaying and the work in rising shall be properly toothed and racked back.

All bricks shall be wetted sufficiently prior to laying to avoid excessive suction. All bed and vertical joints shall be filled solid with mortar as the bricks are laid. Bricks shall be laid with frogs uppermost and shall be wetted during hot weather.

Prices for brickwork shall include the cost of all necessary scaffolding.

prescribe general rules governing bricklaying to ensure thoroughly sound and stable brickwork. Note the reference to the relevant Code of Practice.

Bonding of brickwork

Walls exceeding 102·5 mm (4½ in.) in thickness shall be built in English bond unless otherwise directed by the Engineer. Half-brick walls shall be constructed in stretcher bond. Hollow walls shall be constructed of two half-brick skins separated by a 50 mm (2 in.) cavity and tied with three 150 mm (6 in.) galvanised strip-type fish-tailed ties to B.S. 1243 per square metre (sq. yd) in staggered formation. All ties shall be kept clear of mortar droppings. The use of snap headers shall not be permitted and bats shall be allowed only as closures.

It is necessary to specify the bonds to be used in the construction of the brickwork and to state the number and type of wall ties in hollow walls.

Labours to brickwork

The Contractor shall build in or cut and pin the ends of joists, lintels, steps,

The type of labours involved will vary from job to

92

corbels, etc., and shall build in all frames and bed and point in cement mortar.

The Contractor shall perform all necessary rough and fair cutting and shall form all necessary chases and reveals.

The Contractor shall rake out joints of brickwork for the insertion of metal flashings, wedge the flashings and afterwards repoint the joints in cement mortar.

job and the specification writer must examine the drawings carefully, in order to pick up all the labours needed and describe them in the specification.

Bricklaying in frosty weather

When frost is likely to occur brickwork shall be properly protected and covered with sacking, tarpaulins or other suitable material in the manner recommended in CP 121 Part I. Any brickwork which has been affected by frost shall be pulled down and rebuilt at the Contractor's expense.

Protection shall also be provided against heavy rain or other severe weather conditions.

The Contractor is invariably required to protect new brickwork adequately against frost and other severe weather conditions. Responsibility for the replacement of work damaged by the lack of such protective measures rests with the Contractor.

Damp-proof courses

(a) *Engineering bricks.* The Contractor shall lay over the full thickness of all walls in the positions indicated on the Drawings, a damp-proof course of two courses of class B engineering bricks to B.S. 3921 bedded and pointed in cement mortar (1:3).

(b) *Bitumen felt.* The Contractor shall lay over the full thickness of all walls in the positions indicated on the Drawings a damp-proof course of a single layer of bitumen felt, incorporating a hessian

Various damp-proof course materials are available, from engineering bricks and slates to bitumen felt and sheet metals. Two of the most popular materials have been taken to illustrate the approach usually adopted. Note the reference to relevant British Standards wherever possible and the need to state the extent of laps with materials supplied in rolls.

93

base and a layer of lead in accordance with B.S. 743, type 5 D. Joints in the bitumen felt shall be kept to a minimum and the damp-proof course shall be lapped 225 mm (9 in.) at joints. The bitumen felt shall be bedded and pointed in cement mortar (1:3).

Faced brickwork

Facing bricks of the type specified shall be laid in the positions indicated on the Drawings and all facing brickwork shall be well bonded to the backing bricks. No facing brickwork shall at any time be more than 600 mm (2 ft) above the backing brickwork.

All facing brickwork shall be pointed with a rubbed joint as the work proceeds and internal faces of brickwork shall be pointed with a neat flush joint to give a fair face.

Faced work shall be kept clean at all times and scaffold boards adjoining brick faces shall be turned back at nights or during heavy rain. All faced brickwork shall be cleaned down as necessary on completion to give an even-coloured surface free of mortar droppings or staining of any kind. The Contractor shall carefully fill all putlog holes with bricks similar to the surrounding brickwork, point up as required and generally make good.

Faced brickwork is normally provided to the external wall faces of buildings to enhance their appearance. Numerous types of facing bricks and methods of pointing are available. Pointing can be performed as the bricklaying proceeds or be left until the brickwork is complete. The joints are then raked out and pointed to secure a uniform treatment throughout. Brick joints can be finished flush, struck or weathered; have a shallow, rounded or rectangular recess; or take a more complex form.

Reinforced brickwork

Reinforced brickwork shall be provided with strips of No. 20 gauge expanded metal in each bed joint as shown on the Drawings. The expanded metal

Brickwork is sometimes reinforced to increase its resistance to tensile and shear stresses.

94

reinforcement shall be lapped a minimum of 450 mm (18 in.) at joints.

Chimney shaft linings

Chimney shaft linings shall be formed of a 115 mm (4½ in.) lining of Messrs X or other approved solid grade insulating blocks jointed in mortar composed of one part Portland cement and four parts of Messrs X No. 6 F powder or other approved mix. The linings shall be tied to the main brickwork with galvanised steel ties, as specified, 1 metre (3 ft) apart and in every third course.

Tall chimney shafts often have an inner lining of special heat-resisting bricks. As the lining is not bonded to the main brickwork, it is often considered necessary to insert ties connecting the two.

Clauses covering any other special classes of brickwork, such as glazed bricks and brick arches, can conveniently follow at this stage.

MASONRY

Dressed stonework

All dressed stone shall be sandstone obtained from Messrs X's quarry and shall be free from vents, clayholes, discolourations or other defects and be of even texture and colour. The stone shall be laid on its natural or quarry bed.

All dressed stone is to match in colour and texture the samples of stone submitted to and approved by the Engineer. Every stone shall ring clearly when struck with a hammer.

The mortar for bedding and jointing shall consist of one part ordinary Portland cement to B.S. 12, one part lime and six parts sand. The lime shall be semi-hydraulic lime conforming to B.S. 890, Part I, class B, and the sand shall comply with B.S. 1200 (grading as Table 1).

Dressed stone or ashlar is usually employed as a stone face to a rubble, brick or concrete wall, to produce a first-class finish and possibly to harmonise with walling in the vicinity of the job.

Ashlar is usually defined as masonry consisting of fine blocks of stone, finely square-dressed to given dimensions and laid in courses of not less than 300 mm (12 in.) in height with fine joints.

The specification clauses should preferably begin with material requirements and then follow with details of workmanship.

95

Each stone shall be set on a full mortar bed not less than 5 mm ($\frac{3}{16}$ in.) thick and the beds shall be damped with water before setting the stonework. All stonework shall be carried out in general conformity with CP 121.201.

Each stone shall be of the dimensions shown on the Drawings and the backs shall be dressed at right angles to the bed. All beds shall be set horizontally and dressed to even surfaces throughout. All joints shall be dressed for the full depth of the stone and the exposed faces shall be chiselled to a fair surface.

The whole of the dressed stone shall be set in gauged mortar (1:1:6) and neatly jointed as the work proceeds. Clean down all exposed faces of dressed stone to remove all dirt and mortar stains and leave in perfect condition free from any defacement.

The prices for dressed stone shall include for chiselled exposed faces, preparatory labours, beds, joints, back faces, sunk faces, splays, transporting, storage, hoisting, setting in mortar and jointing as the work proceeds.

No stonework shall proceed when the temperature falls below 1 °C (34 °F) and all work must be adequately protected with sacking, tarpaulins or other suitable material against frost and rain. Any work damaged by frost or rain shall be relaid with fresh mortar at the Contractor's expense.

The Contractor shall suitably protect all dressed stonework and make good any damaged stonework at his own expense and to the satisfaction of the Engineer. Any chipped or broken stones shall be cut out and replaced.

To obtain a satisfactory finish to the stonework, it is necessary for the stones to be carefully dressed and laid with fine joints. Provision should also be made for the protection of the stonework from damage by frost, rain or other causes and for the replacement of damaged work by the Contractor at his own expense.

Mason's mortars are sometimes used in preference to gauged mortars, a common mix by volume being:

12 parts crushed stone

3 parts lime putty or hydrated lime

1 part Portland cement.

Rubble walls

Squared rubble, built to courses. Squared rubble shall consist of approved stone from Messrs X's quarry and no stones shall be less than 225 mm (9 in.) long, 150 mm (6 in.) wide and 100 mm (4 in.) deep. All stones shall be roughly squared and dressed smooth on beds and joints for a distance of at least 100 mm (4 in.) from the exposed face. Bond stones shall be provided at the rate of at least one to each square metre (sq. yd) of exposed face: they shall measure at least 150 mm × 150 mm (6 in. × 6 in.) on the face and extend for three-quarters of the thickness of the wall. Vertical joints shall not extend past more than three stones and the horizontal lapping of stones shall be not less than 100 mm (4 in.). The stonework shall be levelled up at intervals of about 600 mm (24 in.), and all stones shall be solidly bedded and jointed in gauged mortar (1:1:6) with flush joints as the work proceeds.

In squared rubble the stones are roughly squared and small stones are sometimes introduced to assist bonding. These are generally referred to as 'sneck' stones. In work built to courses the stonework is levelled up at intervals varying between 300 mm (12 in.) and 1 m (36 in.) in the height of the wall. This form of walling provides a reasonably economical and regular stone finish. Squared rubble can also be built coursed or uncoursed.

Random rubble, uncoursed. Random rubble shall consist of approved stone from Messrs X's quarry, carefully selected by the mason or waller to obtain a good bond and shall be hammer-pitched on exposed surfaces. Bond stones shall be provided at the rate of at least one to each square yard of exposed face, they shall measure at least 150 mm × 150 mm (6 in. × 6 in.) on the face and extend for three-quarters of the thickness of the wall.

Stones shall be solidly bedded and jointed in gauged mortar (1:1:6) with flush joints as the work proceeds. The

Random rubble may be uncoursed or brought to courses. This represents the cheapest form of stone walling and has an irregular appearance. Bond stones are needed to give the wall adequate strength and the interior voids are usually filled with small stones.

interior spaces in the wall shall be filled with small pieces of stone grouted in position.

Dry rubble walling. Dry rubble boundary walls shall be constructed of approved stones of the type specified for random rubble. The stones shall be roughly dressed and laid on edge at a slope with the hearting formed of small stones. The stones shall be laid in such a manner that rain penetrating the outer face will run out at a lower level.

This type of walling is occasionally used for boundary walls in rural areas and is similar to uncoursed random rubble with the omission of mortar from the joints.

Special stonework

Coping stones in natural stone shall be of approved stone, dressed with a smooth finish to the shapes and dimensions shown on the working drawings, and shall comply with the requirements of B.S. 3798. The coping stones shall be laid to the correct lines and bedded and jointed in gauged mortar (1:1:6).

Quoins, plinths, voussoirs and other special stonework shall be of approved stone dressed to the finishes, shapes and dimensions shown on the working drawings and bedded and jointed in the manner indicated.

This clause refers the Contractor to the working drawings for some of the more detailed requirements. Where this procedure is adopted, it is essential that the Contractor should be supplied with these drawings at the time of tendering. Alternatively, a full description can be given in the specification, such as '375 mm × 150 mm (15 in. × 6 in.) parapet coping, weathered on top with 60 mm ($2\frac{1}{2}$ in.) throated projection at each edge'.

Dowels and cramps

Copper dowels, 25 mm (1 in.) square and 75 mm (3 in.) long, shall be fixed between column stones and in other positions indicated on the Drawings or where directed by the Engineer. The dowels shall penetrate an equal distance

Both dowels and cramps are used to join stones together and keep them in their correct positions. Dowels are fixed in a vertical plane and cramps are used horizontally.

into each stone and shall be set in cement mortar.

Gunmetal cramps, 225 mm (9 in.) long by 40 mm (1½ in.) wide by 5 mm (¼ in.) thick with each end turned at right angles for a distance of 40 mm (1½ in.), shall be provided at each joint between coping stones. The Contractor shall form suitable mortices to receive the cramps which shall be set in cement mortar. The cramps shall extend for an equal distance into each stone.

As an alternative, slate can be used for both dowels and cramps. The metal used must be non-corrodible.

Cast stonework

Cast stone shall be obtained from an approved manufacturer and shall comply with the requirements of B.S. 1217. The facing shall be not less than 20 mm (¾ in.) thick and shall consist of one part of white Portland cement to three parts of Bath stone aggregate with a maximum size of 15 mm (½ in.) The faces of the stones shall be smooth, even and free from crazing, and equal to samples submitted to and approved by the Engineer. Aggregates used in the concrete core shall comply with B.S. 882.

The Contractor shall be responsible for protecting cast stonework on the site. Metal bond ties shall be cast into the stones as shown on the Drawings or directed by the Engineer.

The stones shall be bedded and pointed in one operation with a mason's mortar consisting of one part Portland cement to two parts white hydrated cement and eight parts clean fine aggregate, all by volume. Facing stones shall be brought up in courses to a height not exceeding 1 m (3 ft) in one operation.

Cast stone is a material manufactured from cement and natural aggregate for use in a manner similar to and for the same purposes as natural stone. The term includes reconstructed stone and artificial stone.

Cast stone often consists of a comparatively thin facing incorporating crushed stone, around a core of structural concrete.

Reference should be made to B.S. 1217: Cast Stone, for basic requirements and any additional particulars added in the specification.

The concrete backing shall then be brought up and well compacted around the stones and projecting metal ties.

WATERPROOFING

Asphalt

Asphalt for tanking and damp-proof courses shall comply with the requirements of B.S. 988 (mastic asphalt for building—limestone aggregate) or B.S. 1162 (mastic asphalt for building—natural rock asphalt aggregate).

The most efficient waterproofing membrane consists of asphalt, of which mastic asphalt is the most suitable for this purpose.

Asphalt work

All asphalt work shall be carried out by workmen experienced in the preparation and laying of mastic asphalt.

Horizontal membranes shall be laid in three thicknesses, each with 75 mm (3 in.) laps, to a total thickness of 30 mm ($1\frac{1}{8}$ in.), and shall be carried through walls to connect with vertical membranes with a two-coat angle fillet in the manner shown on the Drawings. All surfaces to be covered with asphalt shall be dry and free from dirt or loose material immediately prior to the application of asphalt.

Vertical membranes shall be applied in three thicknesses with 75 mm (3 in.) laps, to a total thickness of not less than 20 mm ($\frac{3}{4}$ in.).

Where necessary, protect the asphalt by laying loading coats of concrete as soon as each section of asphalt work is complete. Pumping shall be continued until the concrete loading coat is set.

To ensure satisfactory results it is necessary to lay the asphalt in a minimum number of coats with a minimum total thickness and the requirements vary for different situations. The accompanying specification clauses indicate the generally accepted minimum requirements. Other basic requirements include the provision of laps between adjoining coats of asphalt and two-coat angle fillets at the junction of horizontal and vertical work.

Asphalt tanking must be continuous and it should therefore be carried under stanchion bases by lining their pits. It is also important to protect all asphalt as the work proceeds. Where a

Asphalt for use on roofs shall comply with B.S. 988 or B.S. 1162 and shall be laid to falls on an underlay of sheathing felt. Mastic asphalt to flat roofs shall be laid in two thicknesses with 75 mm (3 in.) laps, to a total thickness of not less than 20 mm ($\frac{3}{4}$ in.). At the junctions of flat roofs and parapet walls, two-coat asphalt skirtings shall be provided 150 mm (6 in.) in height above the highest part of the roof. Two-coat angle fillets shall be provided at the junction of the skirting and the roof, and the top of the skirting shall be splayed, turned at least 25 mm (1 in.) into a groove in the brickwork or concrete, and pointed or connected to a damp-proof course in the parapet wall. Mastic asphalt roofing shall be laid in accordance with CP 144 Part 4.

horizontal layer of asphalt is to be covered by reinforced concrete, it is advisable to lay a 50 mm (2 in.) layer of concrete in advance of the laying of the reinforcement to avoid damage to the asphalt.

Bitumen sheeting

Bitumen sheeting shall comply with B.S. 747 type 1C, weighing not less than 1 kg/m^2 ($2\frac{1}{2}$ lb/yd^2). For roof coverings the sheeting shall be applied in three layers, the bottom layer being of asbestos based felt, type 2A, and the top layer shall be covered with white Derbyshire spar chippings set on a coat of cold dressing compound.

All surfaces shall be dry and free from dirt and loose material immediately prior to the application of bitumen primer or an approved cut-back bitumen solution of suitable viscosity. The sheeting shall be laid with laps of not less than 50 mm (2 in.), and a coat of hot oxidised or blown bitumen bonding compound shall be applied between adjacent sheets. Splayed or rounded

An alternative and cheaper form of waterproofing is provided by the use of bitumen sheeting. The sheeting should conform to one of the classes detailed in B.S. 747, it should be properly lapped, and each sheet should be fixed and sealed with hot bitumen to ensure a sound and waterproof job.

101

angle fillets shall be provided at the junction of horizontal and vertical surfaces.

Bitumen felt roof coverings shall be laid in accordance with CP 144 Part 3: Built-up Bitumen Felt. The Contractor shall be responsible for the protection of all bitumen sheeting throughout the period of the Contract.

Chapter Seven

Specification of Piling

Extensive use is made of piles in civil engineering work, often for the purpose of transmitting heavy loads down to a firm stratum at a considerable depth below ground level. Sheet piling is used extensively to hold back water or loose soil and to form the walls of wharves, jetties, etc.

There are three main classes of piling: timber, reinforced concrete and steel sheet. Reinforced concrete piles can be cast or formed in situ, and there is a wide variety of proprietary in situ concrete piling systems. In one method a steel lining tube is sunk by a mechanical auger until a satisfactory bearing stratum is reached, concrete is forced into an enlarged base, steel reinforcement with helical binding is lowered down the temporary lining tube and further concrete is placed and rammed as the lining tube is withdrawn. Another method makes use of compacted gravel in place of concrete and the interstices in the gravel are subsequently filled with cement grout injected under pressure.

Different methods can be employed for excavating the soil from the pile holes, such as the use of cutters or shells inside lining tubes. Yet another system entails the driving of a conical cast iron shoe at the base of a steel tube; the steel tube may be left in position or withdrawn according to circumstances. In both cases steel reinforcement and concrete will be placed in the void. Another system uses a revolving screw pile shoe to bore through the various strata.

When specifying contractor-designed concrete piles it is usual to state the superimposed load to be carried by each pile, the method of disposal of any surplus spoil, any restrictions regarding the type of pile or method of driving and the finished levels of tops of piles in relation to ground level.

With piles other than contractor-designed, it is advisable to set out the specification requirements in a logical sequence to assist the Contractor in reading the document and to reduce the risk of omission of essential details. As with the work previously described, it is good practice to insert materials clauses for each class of piling followed by workmanship and other requirements.

The following lists of items give a typical range of clauses covering each of the three main classes of piling.

Concrete Piles (cast)

  (1) Concrete
  (2) Reinforcement
  (3) Shoes
  (4) Casting
  (5) Curing, stripping and stacking
  (6) Ready-made piles
  (7) Trial piles
  (8) Handling
  (9) Pitching and driving
(10) Lengthening piles

Alternatively, concrete piles may be formed in situ, when they are likely to be constructed under a proprietary system. When engineer-designed, it is usual to specify the dimensions of piles, class of concrete and type and size of reinforcement. In addition, clauses normally will be included relating to compaction of concrete and cover to reinforcing bars, and the Contractor is usually required to keep a record of the construction of each pile and to test certain piles.

Timber Piles

(1) Timber
(2) Piles
(3) Creosoting or tarring
(4) Shoes and rings
(5) Pitching and driving
(6) Cutting off heads of piles

STEEL SHEET PILING

(1) Steel sheet piles
(2) Driving
(3) Damaged piles
(4) Cutting piles
(5) Drilling piles

A selection of typical specification clauses for piling follow:

TYPICAL SPECIFICATION CLAUSES EXPLANATORY NOTES

CONCRETE PILES (CAST)

Concrete

Concrete shall be mixed in the proportions of 50 kg (112 lb) of Portland cement to 0·13 m³ (4½ ft³) of aggregate with a maximum size of 20 mm (¾ in.). Minimum compressive strengths 28 days after mixing shall be 39 MN/m² (5600 lbf/in.²) on a preliminary test and 29 MN/m² (4200 lbf/in.²) on a works test.

The cement and aggregate shall comply with the requirements under 'Concrete Work' (see Chapter 5).

Mix and strength requirements should be given, but the Contractor can be referred to previous clauses for materials, and for gauging and mixing requirements as necessary. In some cases rapid-hardening cement may be specified. Alternatively, the compressive strengths might be expressed in N/mm.²

Reinforcement

Reinforcement shall consist of 30 mm (1¼ in.) diameter mild steel bars and 6 mm (¼ in.) diameter binding links to B.S. 785, bent and fixed as shown on the Drawings. Main reinforcing bars shall be supplied in one complete length as far as possible; where this is impracticable, separate lengths shall be satisfactorily spliced or butt welded. Steel skeletons or cages shall be fabricated before being placed in the moulds.

General requirements with regard to the nature and method of fixing steel reinforcement should be given. Note the requirements relating to fabrication of steel cages and the use of forks or stretchers.

105

Pressed steel forks or stretchers, machine moulded and true to length, shall be fixed in the positions indicated on the Drawings, to keep the binding links taut.

Shoes

Chilled hardened cast iron shoes with mild steel straps, weighing 25 kg (56 lb) each, as supplied by Messrs X or other equal and approved, shall be provided, shall be accurately fitted to the piles and tied to the pile reinforcement.

The type and weight of shoe should be stated, together with fixing requirements.

Casting

Concrete piles shall be cast in wrought timber or other approved moulds on a substantial horizontal platform capable of carrying the weight of the piles without appreciable deflection. The concreting of each pile shall be completed in one continuous operation free from interruptions of any kind.

All arrises of piles shall be chamfered and holes for lifting tackle shall be formed through piles in the positions shown on the Drawings.

General requirements are given as to moulds and casting platforms. Timber, steel or composite moulds would normally be permissible. It is essential that each pile should be cast in a single operation. Note the requirement as to chamfered arrises.

Curing, stripping and stacking

Piles made with ordinary Portland cement shall be kept moist for a minimum period of 14 days after casting. Side forms may be stripped 1 day and bottom forms 10 days after casting, provided the piles are kept supported on level blocks spaced not more than 2 m (6 ft) apart.

It is essential that piles should be properly cured and that forms are not stripped prematurely. Piles must not be driven before they have matured sufficiently. The periods quoted can be reduced where rapid-hardening

106

Piles may be lifted 21 days after casting and moved to a suitable stacking site, but they shall not be driven until at least 6 weeks after casting. Each pile shall be suitably marked with the date of casting and the stacks so arranged as to permit the use of piles in correct age order.

and other special types of cement are used. The DOE Specification for Road and Bridge Works specifies a minimum period of 28 days between casting and driving.

Ready-made piles

The Contractor may use ready-made piles with the approval of the Engineer. The Contractor shall provide details of the manufacturer and the piles to be supplied, which shall be produced strictly in accordance with the requirements of the previous clauses. Care shall be taken to prevent damage to the piles in transit.

The use of ready-made piles will often result in the earlier supply of piles, but care is needed to prevent damage to piles in transit. The Engineer will have right of access to the supplier's casting yard.

Trial piles

The Contractor shall, as soon as practicable after the acceptance of his tender, cast, drive and test trial piles made with an approved rapid-hardening cement, either singly or in groups in permanent positions as directed by the Engineer. The trial piles shall be manufactured in conformity with the requirements of the relevant clauses in the specification.

It is customary for the Contractor to be required to cast, drive and test a number of trial piles on the site with a view to determining the probable lengths of the piles generally.

Handling

Piles shall be carefully handled and lifted and shall not be jolted or stacked in a manner which will subject them to bending. No pile shall be lifted other than by slinging from the lifting holes.

The lifting holes are usually formed at the quarter points in the pile. Alternatively, cast iron pipes may be cast into the piles for lifting purposes.

Pitching and driving

Piles shall be pitched accurately in the required positions and shall be driven to the lines shown on the Drawings. Any piles which are out c ° alignment or verticality shall be withdrawn and re-pitched.

Single acting rams or drop hammers weighing not less than 3 tonnes (3 tons) and giving 40 blows per minute with a drop of 1 m (3 ft) shall be used for driving piles. Driving shall continue until the piles penetrate to a minimum depth of 7·5 m (25 ft) below dredged level or until a set of 25 mm (1 in.) has been obtained for the last 10 blows of the monkey with the specified drop.

Raking piles shall be driven accurately to the rake shown on the Drawings. The pile frame shall be fitted with leads capable of adjustment to the required angle and extension leads shall be used where necessary.

Piling frames must be capable of driving piles below the level of the base of the frame. The Engineer or his representative shall be present when each pile is driven to its final set. Suitable helmets shall be fitted to the heads of piles to prevent damage during driving.

The Contractor shall supply the Engineer with a report containing a complete record of the results of the driving of each pile.

After piles have been driven to the required set and to the satisfaction of the Engineer, the concrete shall be cut away from the head of each pile for a distance of 600 mm (2 ft) and the main reinforcing bars bent as necessary for connection to intersecting members.

It is usual to indicate in the specification the type of piling plant to be used and the method of determining the lengths of piles. Other common requirements cover the use of helmets, stripping of heads of piles and supply of records covering pile-driving operations.

On occasions the lengths of piles are determined by reference to test piles driven in advance of the main piles. The Engineer will then decide on the lengths to which the remainder of the piles are to be constructed and may vary the type and weight of hammer to be used.

The weights of hammers vary between 2 and 4 tonnes (2 and 4 tons) and the drop or free fall of the hammer is often 1 m (3 ft or 3 ft 6 in.).

The Contractor's report normally gives details of pile reference and location, and shortening or lengthening details, depth of penetration, set and safe load.

The DOE Specification for Road and Bridge Works requires the Contractor to keep a record of all piles driven.

Where piles extend for more than 600 mm (2 ft) above the required level, the surplus length shall be cut off and removed unless the Engineer directs otherwise. The Contractor's billed rates shall include for stripping heads and cutting off surplus lengths of piles.

Lengthening of piles

Where piles require to be lengthened, the reinforcing bars shall be stripped of concrete for a minimum distance of 600 mm (2 ft). The old concrete must be adequately roughened and rinsed with clean water prior to the application of a neat cement slurry and a 5 mm ($\frac{1}{4}$ in.) thick layer of cement mortar of the same proportions as that contained in the concrete mix.

The additional reinforcement shall be spliced or butt-welded to the exposed reinforcing bars, as directed by the Engineer. The new concrete shall be of the same mix as the original concrete and shall be adequately compacted between suitable moulds. The Engineer may permit the use of rapid-hardening Portland cement or high alumina cement in the new concrete where desirable.

The specification requirements aim at securing a satisfactory connection between the new and the old concrete and reinforcement respectively. It is customary to use butt joints for this purpose. The use of rapid-hardening or high alumina cement may be permitted to speed up the work.

TIMBER PILES

Timber

Timber for piles shall be greenheart of good quality, straight, sound, sawn square, well seasoned and free from rot, worm, beetle, injuries, shakes, large and decayed knots or other defects and shall conform to B.S. 1860.

It is good policy to begin with a clause covering timber requirements to ensure the use of timber of adequate strength and durability. Note the reference to B.S. 1860 dealing with the measurement of characteristics affecting the strength of structural softwood.

Piles

Piles shall be at least 300 mm × 300 mm (12 in. × 12 in.) and shall be properly pointed and shod, and be provided with rings at their heads. Piles are to be driven at least 3 m (10 ft) into the river bed.

A common size for timber piles is 300 mm × 300 mm (12 in. × 12 in.). Average lengths are sometimes included in the specification particulars.

Creosoting

Creosoting of timber piles shall be carried out in accordance with the requirements of B.S. 913 (full cell process) and the creosote shall comply with B.S. 144. The timber shall be worked, incised 20 mm (¾ in.) deep and have clean surfaces and a moisture content not exceeding 25 per cent prior to pressure creosoting. All later cuts shall receive two brush coats of creosote prior to the fixing of the timber in its final position.

It is essential to list preparation work on the timber prior to creosoting and to detail the method of creosoting. Note the use of British Standards to ensure a good standard with a minimum of description.

Tarring

All connecting surfaces of timbers, scarf joints and cut ends shall receive two coats of hot coal tar complying with B.S. 3051 before the timbers are fixed. After completion the whole of the timberwork shall be painted with two coats of hot coal tar.

The application of coal tar is an alternative to creosote. The Contractor is referred to a British Standard for the detailed requirements.

Shoes and rings

The points of all timber piles shall be protected with cast iron shoes weighing

Timber piles need protection by metalwork at both

not less than 12·5 kg (28 lb) each, securely fixed to the timber with wrought iron straps. The heads of timber piles shall be protected with tightly fitting mild steel or wrought iron rings, 75 mm (3 in.) wide × 20 mm (¾ in.) thick.

ends. Sufficient information should be given to avoid any doubt as to the Engineer's requirements.

Pitching and driving

Timber piles shall be accurately pitched in the required positions and driven to the lines and levels shown on the Drawings or as directed by the Engineer. The driving hammer shall weigh not less than 2 tonnes (2 tons), shall not drop a greater distance than 2 m (6 ft), and shall be guided by leads to ensure that the piles are driven in correct alignment and to the required batter.

Any piles driven out of line, broken, split or otherwise damaged are to be withdrawn and replaced where necessary. Maximum permissible deviations shall be 50 mm (2 in.) for alignment and 2 per cent for verticality.

The depth of penetration of the piles may be determined in various ways, including the attainment of a certain set when the length of pile is not readily determinable. The weight of the driving hammer may be varied to suit the particular strata encountered.

Timber piles to wharves and jetties are often driven to a batter of about 1 in 24.

Cutting off pile heads

After timber piles have been driven to the required set and to the satisfaction of the Engineer, the heads shall be cut off square at the levels shown on the Drawings or as directed by the Engineer. When estimating the lengths of timber piles the Contractor shall make allowance for the removal of damaged timber in pile heads.

Even although rings are provided, timber in the heads of piles becomes compressed and splits from the impact of the driving hammer. It must then be removed.

111

STEEL SHEET PILING

Steel sheet piles

Steel sheet piling shall be Messrs X No. 3 section, weighing 57 kg/m (41·35 lb/ft) or other equal and approved, and it shall conform with the requirements of B.S. 4360: Weldable Structural Steels. The piles shall be free of pronounced warp, have properly formed interlocks and be free from cracks at folds. The piling shall be coated with an acid-resisting tar based paint before delivery to site.

The Contractor shall provide all special closer piles and shall allow for thoroughly greasing the locks of piles before pitching.

Driving

The Contractor shall provide all necessary frames, leaders, etc., needed for driving the steel sheet piling. A suitable helmet shall be provided to prevent damage to the head of the pile when driving. The piles shall be guided and held in position by adequate temporary walings and struts and all precautions shall be taken to ensure that the piles are driven in correct alignment. The Contractor shall submit to the Engineer or his representative for approval the method he proposes to use for the driving of piles.

The sheet piles shall be driven accurately and truly vertical to the lines and levels shown on the Drawings or as directed by the Engineer. Any creep occurring shall be eliminated in a manner approved by the Engineer.

The Engineer may specify a particular make and type of pile or specify that the piles shall be of approved type, giving greater flexibility to the Contractor.

The Contractor is required to provide all plant, equipment and labour necessary to drive the steel sheet piling satisfactorily in the required positions and to the appropriate depths.

The piles may be driven singly or in pairs and they must be properly interlocked throughout the whole of the driving period. It is good practice to specify the action to be taken when obstructions are encountered or when piles are driven below the required level.

Where the steel sheet piling is to be tied back to

112

Where the driving of any pile or pair of piles meets obstruction or resistance, it shall be discontinued and the driving of other piles commenced and/or continued until the final level is reached. The Contractor shall return to the piles meeting obstruction or resistance and recommence driving until the final level is reached or, on the instructions of the Engineer, the affected piles are cut off at a suitable level.

Any piles driven deeper than the required level shall be withdrawn to the correct level. Existing piles on the site of the new work shall be withdrawn completely and re-driven in new positions as shown on the Drawings.

concrete, one method is to weld mild steel tangs, about 15 mm × 50 mm × 375 mm ($\frac{1}{2}$ in. × 2 in. × 15 in.) girth, to the back of the sheeting. One end of each tang will probably be split and fishtailed.

Damaged or misplaced piles

Any piles which are driven out of alignment or twisted, broken, bent or otherwise damaged in driving shall be withdrawn and replaced, at the Contractor's expense, with other piles which are properly driven to the satisfaction of the Engineer's representative.

This clause provides for the removal of unsatisfactory piles and their replacement with sound, properly driven piles.

Cutting piles

The Contractor will not be permitted to cut steel sheet piling, except where shown on the Drawings or as ordered by the Engineer. Any cutting shall be performed with an approved type of oxy-acetylene burning plant.

Cutting of steel sheet piling is generally restricted as to extent and method.

Drilling piles

The Contractor shall drill the steel sheet piling as necessary and shall fix steel channel waling and tie rods as shown on the Drawings.

This is a reminder to the Contractor that his piling price must include for the additional labours involved.

CHAPTER EIGHT

Specification of Iron and Steelwork

THIS chapter is primarily concerned with the drafting of specification clauses covering structural steelwork with riveted and welded connections. Incidental items of metalwork such as ladders, handrailing and open steel flooring, and some of the more common forms of wall and roof sheeting are also included.

Probable main headings in a structural steelwork specification are as follows

(1) Structural steel
(2) Fabrication
(3) Inspection and marking
(4) Erection
(5) Bolting
(6) Riveting
(7) Welding
(8) Measurement
(9) Testing
(10) Painting

Structural steelwork is normally supplied and erected by a steel fabricator, who may also carry out the painting of the steelwork. On very large structural steelwork contracts the steel fabricator, as main contractor under the steelwork contract, may also be responsible for incidental work such as roof decking, ventilators and patent glazed lanterns, which are to be provided by nominated sub-contractors. The steelwork contractor is usually required to prepare working drawings of the permanent steelwork and to submit them to the Engineer for approval before any work is put in hand. He also furnishes corrected or amended drawings as necessary.

STRUCTURAL STEELWORK

Structural steel

Structural steel shall be to the approval of the Engineer or his representative and shall comply with the requirements of B.S. 4360: Weldable Structural Steels. Steelwork for bridges shall comply with B.S. 153: Steel Girder Bridges.

Steel castings shall comply with the requirements of B.S. 3100: Steel Castings for General Engineering Purposes.

Numerous references are made to British Standards which lay down stringent standards and tests to be performed on the steel. For instance, B.S. 4360 specifies the process of manufacture, chemical composition, tolerances, and describes the tests to be carried out on plates, sections and bars. The standard also covers selection, certificates and testing procedures.

Fabrication

Workmanship and general fabrication procedure shall be in accordance with B.S. 153, Part 1, where appropriate, and with the best modern practice for structural steelwork.

The edges of universal plates or flats need not be machined except for accurate fitting against adjacent parts. All butting members shall have their ends machined after fabrication.

Where turned bolts are to be used, the holes shall be reamed through the full thickness while the members are held in their correct relative positions. With the approval of the Engineer turned bolts may be used as an alternative to rivets for site connections.

Most of the normal fabrication workmanship requirements are contained in B.S. 153, Part 1. Any additional requirements should be inserted in the specification.

Members can be connected in several different ways: bolting, riveting, welding and, more recently, by the use of high strength friction grip bolts (B.S. 3139 and 3294).

In connection with girder work, it is sometimes specified that plates and bars shall be accurately assembled, and shall be of uniform thickness

115

All stiffeners shall bear tightly against the compression flange and against the loaded flange at points of concentrated load. At other points the ends of riveted stiffeners shall, unless tightly fitted, provide a clearance of at least 10 mm ($\frac{3}{8}$ in.) between their ends and the tension flange, with the outstanding leg bevelled. The ends of welded stiffeners shall be sawn square or to the correct bevel and shall fit tightly against both flanges.

All members shall be to the dimensions shown on the Drawings, cut to exact lengths and finished true and square.

All holes shall be accurately marked from templates and shall be drilled to give smooth edges. The diameters of holes shall not exceed those of the bolts or rivets by more than 1·5 mm ($\frac{1}{16}$ in.).

The Engineer may require the Contractor to erect each part of the steel structure temporarily in the shop, to check the accuracy of the work.

and free from winding, with the surface of adjoining plates and bars in close contact.

Another requirement might be that plates shall be of sufficient size to permit 3 mm ($\frac{1}{8}$ in.) to be planed off all edges.

With regard to bolt and rivet holes, it will be noted that punching of holes and drifting will not be permitted.

Inspection and marking

The Contractor shall notify the Engineer when materials are ready for inspection at the maker's works and fabricated material at the fabricator's works.

After checking or testing at the fabricator's works all members and fittings shall, for the purpose of identification during erection, have a distinguishing number and letter (corresponding to the distinguishing number and letter on an approved drawing) painted and where possible also stamped on in two positions.

The Contractor must provide adequate facilities for inspection of materials by the Engineer at fabricating and manufacturer's shops. It is good practice to have all structural members and fittings suitably marked for ease of working on the job.

Erection

Ordinary steelwork may have a tolerance not exceeding 1 in 500 and steelwork around lift shafts shall not be greater than 1 in 1000 from plumb. Girders shall have a camber of uniform curvature of 5 mm for every metre of span ($\frac{1}{16}$ in. for every foot of span).

All columns and bases shall be set accurately to the required lines and levels, and all holding-down bolts shall be strictly in accordance with the detailed drawings. Concrete bases shall be laid not less than one month before the steelwork is placed in position.

Grillages shall be set truly level by careful levelling at each corner. Girders shall be lowered slowly onto their seating cleats, with each end secured initially by at least one bolt and nut.

No member of the structure shall be finally bolted, riveted or welded until the whole or a major section is approved by the Engineer for line, levels and verticality. Connections shall be completed as soon as possible after receipt of the Engineer's written approval, with care being taken not to interfere with existing steelwork in any way.

It is essential that all the structural members should be accurately fixed in the correct positions, and if any tolerances are permitted, then these must be detailed in the specification.

It is advisable to fix individual members temporarily in the first instance and to carry out the final connections when each structure or section thereof has been inspected and approved by the Engineer.

Bolting

All bolts, nuts, rods, straps and the like shall comply with the latest appropriate British Standard. Unless otherwise stated all bolts shall have washers 5 mm ($\frac{3}{16}$ in.) thick, and heads and nuts shall be well forged hexagonal Whitworth screws. All bolts shall be screwed tight with at least one clear thread projecting beyond the nut when tightened

General requirements as to bolts and washers should be supplied where bolting is permitted. It is imperative that all bolts should be screwed up tight and have a projecting thread beyond the nut. This is subsequently riveted down or welded.

117

up, and the projecting thread shall be riveted down or welded.

Tapered washers shall be provided to bolts which pass through the flanges of rolled joists, etc. The external diameter of washers shall be 2½ times the diameter of the bolt.

B.S. 3139 deals with dimensions, mechanical properties and tests of high strength friction grip bolts.

Riveting

Rivets shall be of the size and to the pitch shown on the Drawings, and shall be of best quality mild steel and be set up by hydraulic power to fill the holes completely when closed up. They shall have hemispherical heads with a projection in all directions of not less than ¼ times the diameter of the rivet.

Any rivets which are away from adjoining surfaces, badly formed, cracked or in any way defective, shall be cut out and replaced. No riveting shall be carried out until the work has been approved by the Engineer and the members shall be bolted together in advance of the riveting.

It is important that rivets should be provided of the sizes and in the positions required, and that they should be properly formed with well shaped and adequately sized heads.

Welding

Welding shall be performed by an electric arc process conforming to best British practice and complying with the requirements of B.S. 1856: General Requirements for the Metal-Arc Welding of Mild Steel. All welding shall be carried out by fully trained and experienced welders.

The welding procedure for making each joint shall be approved by the Engineer before work is commenced and the Contractor shall make such trial

The specification requirements for welding need to be given in considerable detail to ensure a good class, sound job. As on previous occasions the use of British Standards can help considerably in this respect.

It is usual to require all welding to be performed by the electric arc process, although it can be carried out

welds as may be required to demonstrate the soundness of the proposed method and the competence of his workmen.

Electrodes shall be grade A, of the best heavy-coated type, and shall be kept in a dry store in unbroken packets. They shall comply with the requirements of B.S. 639: Covered Electrodes for Manual Metal-arc Welding of Mild and Medium Tensile Steel. Fillet welds shall be made with electrodes not less than 6 mm ($\frac{1}{4}$ in.) in diameter.

The welding plant shall be of modern design and of adequate capacity to produce the required current to each welding point without appreciable fluctuations.

All parts to be welded shall be accurately prepared so that they will fit closely together. After assembly and before the general welding is commenced, the parts shall be tack welded with small fillet welds about 50 mm (2 in.) long made with a 5 mm ($\frac{3}{16}$ in.) diameter electrode and a high current. The tack welds shall be of the same quality and size as the first run of the main weld. When the latter is deposited it shall fuse completely with the ends of the tack welds to form a final profile free from irregularities.

either manually with coated electrodes or by a suitable automatic process. When welding high tensile steel it is often necessary to use electrodes of the basic-coated or hydrogen-free variety.

It is necessary to make suitable allowance in the lengths of steel parts for contraction during welding, in order that the finished lengths will be within the normally accepted limits.

Measurement

Prices for steelwork shall include for the supply, fabrication, delivery and erection of the steelwork as shown on the Drawings and to the satisfaction of the Engineer. Steelwork rates shall include the cost of all necessary staging,

Steelwork rates are to include everything required in the supply, manufacture and erection of the steelwork complete. The actual weights of steel sections may vary

plant, equipment and materials required in the erection of the steelwork.

Payment shall be made on the basis of the calculated weights as determined from the dimensions given on the Drawings. Where the weight of any rolled steel is short of the calculated weight by more than $2\frac{1}{2}$ per cent, but the material is nevertheless accepted, then payment shall be made only for the actual weight of steel supplied.

In the case of mild steel plates, the calculated weight shall be based on 185 kg/m², 25 mm thick (40·8 lb/ft², 1 in. thick), and with mild steel standard sections the calculated weight shall be based on the weight per linear metre (foot) specified in the relevant British Standard Specification. No additional weight for weld metal deposited will be included in the measurements.

from the calculated weights due to wear and tear on the manufacturer's plant. The permissible allowance for rolling margin is usually $2\frac{1}{2}$ per cent up or down from the calculated weights. Where a deficiency in excess of $2\frac{1}{2}$ per cent occurs it is customary to pay only on the basis of the actual weights supplied, otherwise the basis of payment is that of calculated weights.

Testing

The Contractor shall test all welds in main plates and such other welds as the Engineer may direct, with X-ray or gamma-ray apparatus.

No part of the steelwork shall at any time be loaded in excess of the designed working load. On completion of steel-framed bridges, the Contractor shall provide, position and move from place to place, such rollers and loaded trailers as the Engineer may require for testing purposes. The Contractor shall also supply and fix, as directed by the Engineer, sufficient instruments for the measurement of deflection and stresses. Any work shown to be defective by these tests shall be removed and replaced at the Contractor's expense.

Adequate testing of structural frameworks to design loadings is most desirable. The responsibility for failure may not always lie with the Contractor, as for instance where the failure results from errors in a design prepared by the Engineer. It was a common practice in some quarters in years gone by to require the Engineer responsible for the design of a structure to remain beneath it while test loads were applied.

MISCELLANEOUS ITEMS

Ladders

All ladders shall be 375 mm (15 in.) wide overall, fabricated of 65 mm × 15 mm (2½ in. × ½ in.) steel strings and stays, with 20 mm (¾ in.) diameter rungs at 225 mm (9 in.) centres shouldered and riveted to the strings. The bottom ends of strings shall be bent 90° and built into concrete, and the top ends shall be turned over to 150 mm (6 in.) radius, returned 450 mm (18 in.) and bolted to precast concrete copings. Prices shall include for rustproof bolt fixings and mortices or drilling in concrete and grouting in. Ladders shall be heavily galvanised after erection.

This clause gives essential details such as materials, dimensions of members, spacing of rungs and finish to top and bottom ends of strings.

Prices should include for fixing and for all protective treatment.

Widths of ladders are sometimes increased to 450 mm (18 in.), and they may be constructed of wrought iron. Alternative protective treatments include applications of epoxy resin or bituminous paints.

Guardrails

Guardrails shall be supported on galvanised steel double ball pattern forged steel standards, spaced at no more than 1·5 m (5 ft) centres, drilled to receive two 30 mm (1¼ in.) bore tubular handrails with the top handrail at a height of 1 m (3 ft) above base. The standards shall be supplied with extensions and rectangular plates for bolting to sides of concrete walls or have flanged bases with ragged shanks for building into concrete. Plates shall be fixed to concrete walls with two 15 mm × 100 mm (½ in. × 4 in.) galvanised rawl-bolts. Prices shall also include for forming mortices in concrete to receive ragged shanks to standards and for grouting them in.

The majority of guardrails are formed of galvanised steel standards and tubular handrails, to a total height of 1 m (3 ft or 3 ft 6 in.). The handrail tubing can be conveniently provided in accordance with B.S. 1387: Steel Tubes and Tubulars suitable for screwing to B.S. 21 Pipe Threads.

The finish to the bases of standards will vary according to whether they are to be connected to the tops or sides of walls.

Handrails shall be made of 30 mm (1¼ in.) nominal bore, galvanised steel tubes to B.S. 1387, heavy grade, with screwed and socketed joints, in lengths as required. Made bends and capped ends shall be provided as necessary.

Open steel flooring

The Contractor shall supply and place in position in prepared rebates Messrs X, type B, or other equal and approved 25 mm (1 in.) deep open steel floor panels with a heavily galvanised finish. Steel angle curbs 50 mm × 50 mm × 5 mm (2 in. × 2 in. × ¼ in.) shall be cast into the concrete at the tops of walls, where shown on the Drawings, with angles formed of welded mitres.

The Engineer often has a particular type of flooring in mind, although even then the Contractor should be given the opportunity to offer an alternative for considera-tion. Chequer plating is usually galvanised with a thickness of 10 mm (⅜ in.).

SURFACE PROTECTION

Painting

(a) *General requirements.* All work required to be painted shall be properly cleaned and rubbed down between each coat. No coat of paint shall be applied until the Engineer has passed the previ-ous coat as dry, hard and entirely satis-factory.

No paint shall be applied on wet surfaces, or in damp or frosty weather. All paintwork shall be undertaken by skilled tradesmen experienced in this class of work.

Welds and adjacent parent metal shall not be painted prior to inspection and approval.

It is essential that all steel surfaces which will be in con-tact after assembly should be thoroughly cleaned and given one or two coats of suitable priming paint. One of the best primers for steelwork is red lead paint, although sprayed metal coatings complying with B.S. 2569 can be most effective.

Steps must be taken to ensure that suitable paints are applied under satisfac-tory conditions to properly

(b) *Works painting*. Before leaving the Manufacturer's works, all surfaces which will, after erection on the site, be in contact with steel or concrete or be otherwise inaccessible, shall be thoroughly cleaned, scraped, wire-brushed and freed from dirt, rust and scale and shall be given two substantial coats of genuine red lead paint of approved quality and composition. All other surfaces shall be left bare. Red lead shall comply with B.S. 217.

(c) *Site painting*. All exposed surfaces of steelwork shall be thoroughly cleaned, as previously specified, and shall as soon as practicable thereafter be given one good coat of suitable red lead priming paint. The cleaning and application of priming paint shall be carried out after the erection of the steelwork on the site and in such sections as the Engineer may approve or consider necessary to arrest or prevent undue corrosion of the steelwork.

The steelwork shall then be painted with two coats of an approved ready-mixed oil undercoat and one finishing coat of approved ready-mixed oil gloss paint. All paint shall be delivered to the site in the manufacturers' sealed containers. The colours of the undercoats and finishing coat shall be to the approval of the Engineer. The total thickness of paint film shall not be less than 0.15 mm ($\frac{6}{1000}$ in.)

Galvanising

Steel and iron work required to be galvanised shall be pickled in dilute muriatic acid, and then stored and

prepared surfaces. British Standards 2521, 2523, and 2525–2527 cover oil-based paints for protective purposes. It is sometimes advocated that steel surfaces which will become encased in concrete should have a cement wash applied to them and that steelwork which is to be in permanent contact with soil, brickwork or masonry should be protected by bituminous paint.

Site painting may be undertaken by the steel fabricator or by the main civil engineering contractor. Where the paint is applied by the steel fabricator the cost is included in the steelwork rates; otherwise the main contractor will be paid in accordance with the normal rules for the measurement of painting work.

Galvanised work is specified, particularly for handrails and standards. Specifi-

123

dipped in a bath of pure virgin spelter, the quality of which shall comply with the requirements of B.S. 3436: Ingot Zinc. All items shall be passed rapidly through the bath, which shall be of sufficient size to take the articles without need for bending. The galvanised articles shall then be washed and brushed.

Galvanising shall be undertaken after corrugating, chipping, trimming, filing and fitting are completed. All galvanised articles shall be covered evenly on all sides and the additional weight after galvanising shall be not less than 0.35 kg/m^2 ($1\frac{1}{4}$ oz/ft^2) of surface galvanised. All galvanised surfaces shall have a bright face with a crystalline structure and all edges shall be clean and free from drops of spelter.

cation requirements such as those listed are necessary to ensure that a sound galvanising process is adopted to produce a first-class finish.

WALL AND ROOF SHEET COVERINGS

Asbestos cement

Wall cladding shall be of coloured asbestos cement panels, 5 mm ($\frac{1}{4}$ in.) thick with an overall depth of about 55 mm ($2\frac{1}{4}$ in.), fixed with end laps of 150 mm (6 in.) and side laps of 75 mm (3 in.) to steel angles at 2 m (6 ft) centres, with 10 mm ($\frac{3}{8}$ in.) diameter galvanised hook bolts and bituminous washers.

Colours available include grey, blue, red, russet, browns and greens. Various types of sheets with varying profiles are available for wall and roof coverings: roof sheets are fixed in a similar manner to purlins.

The desirable characteristics of asbestos cement decking are detailed in B.S. 3717.

Aluminium

Wall cladding shall be of 22 gauge corrugated aluminium sheets, with an

There are many proprietary forms of aluminium

124

average depth of 20 mm ($\frac{3}{4}$ in.) and corrugations to a nominal pitch of 75 mm (3 in.), fixed as for asbestos cement sheets with side laps of 1$\frac{1}{2}$ corrugations and end laps not less than 150 mm (6 in.).

sheeting available, and many types of roof decking, of which aluminium troughing covered with insulation board or wood wool slabs, finished with built-up bitumen roofing, is very popular.

Corrugated steel

Wall cladding shall be of corrugated galvanised mild steel sheeting of 22 s.w.g. with laps and method of fixing as for aluminium sheeting.

This is a less popular alternative which needs painting at regular intervals. It is advisable to apply a coating of oxide of zinc where the sheeting will be exposed to sea air or acid vapour.

Newer materials for roof decking include glass-fibre polyester resin.

125

CHAPTER NINE

Specification of Timberwork

THIS chapter is primarily concerned with the specifying of the large structural timbers used in the construction of jetties, wharves and similar structures, and of their associated labours. Civil engineering jobs do, on occasion, contain subsidiary items of timberwork, such as scumboards on sewage disposal works, wooden steps and footbridges. Some contracts include a small amount of building work, such as pumping stations which can incorporate a number of joinery items like windows and doors.

The type, form and extent of specification clauses relating to timberwork will accordingly vary considerably from job to job. Nevertheless, in all cases a logical sequence of items should be secured. For instance, with a timber wharf or jetty the following specification clause headings would probably be appropriate:

(1) Quality of timber
(2) Workmanship generally
(3) Fender piles
(4) Rubbing pieces
(5) Walings and braces
(6) Guardrails, decking, etc.
(7) Tarring or creosoting
(8) Bolts, etc.
(9) Measurement
(10) Equipment

Subsidiary carpentry items are best taken under appropriate headings covering the function of the timberwork, e.g. steps or footbridge.

Joinery work could be conveniently covered under the following headings:

126

(1) Quality of timber
(2) Workmanship
(3) Windows
(4) Doors
(5) Miscellaneous work
(6) Painting

The typical joinery specification clauses will be kept as brief as possible, as this is essentially building work. A selection of typical timberwork specification clauses follows.

TYPICAL SPECIFICATION CLAUSES   EXPLANATORY NOTES

TIMBERWORK IN WHARVES AND JETTIES

Quality of timber

The timber for fender piles, walings and braces shall be greenheart, rubbing pieces shall be of elm, and Columbian pine shall be used for all other timbers. Timbers in elm shall be creosoted under pressure to absorb 180 kg/m³ (10 lb/ft³), while all Columbian pine timbers shall absorb 125 kg/m³ (7 lb/ft³) of creosote.

All timber shall be of merchantable grade and shall be straight, sound, square cut and free from injuries, waney edges, decay, shakes, large and dead knots, insect attack and other serious defects. Any baulks of timber showing more than 15 per cent sapwood on one end section or more than 10 per cent on average of both end sections shall not be accepted.

All timbers shall be of the scantlings shown on the Drawings, with an allowance of 3 mm ($\frac{1}{8}$ in.) for each wrot face. The moisture content of timbers shall not exceed 22 per cent of the dry weight at time of use. Softwood shall, unless otherwise specified, comply with B.S. Code of Practice 112 and the measur-

The type and grade of timber required must be stated as precisely as possible. Timbers must be free from all defects which would impair their suitability for the function which they have to perform.

Adequate seasoning of timber can be ensured by specifying the maximum permissible moisture content.

Note the use of a British Standard and a Code of Practice: The Structural Use of Timber to assist in ensuring the use of timbers of satisfactory quality.

127

able characteristics and moisture content shall be assessed in accordance with B.S. 1860: Structural Timber: Measurement of Characteristics affecting Strength.

Workmanship generally

Fender piles shall each be in one piece and no joints will be permitted. Joints in walings and braces shall be properly scarfed for a minimum length of 450 mm (18 in.).

All labours on timbers shall be accurately executed and finished in a first-class manner. All tenons, mortices, scarves, rebates and other joints shall be accurately cut and well fitted together in accordance with the best class of workmanship with mild steel bolts, coachscrews, straps, nails, oak pins, wedges, etc., as shown on the Drawings or as directed by the Engineer.

Holes for bolts, etc., shall be accurately drilled in the required positions and be of the correct sizes to secure a tight fit. The holes shall be countersunk where shown to receive the heads of bolts, nuts, etc.

The final completed timber shall be in full accordance with the details shown on the Drawings, with wrought faces where required.

Prices shall include for any nails or screws required for fixing purposes and these shall comply with B.S. 1202 and B.S. 1210 respectively.

These clauses are concerned primarily with the jointing of the structural members. Failure of a timber structure often results from poorly constructed joints. Load-bearing timbers are normally extended by means of scarfed joints, when the tapered ends of adjoining lengths of timber are bolted together. On occasions the lengthening of certain timbers, such as piles, is not permitted.

Intersections of main structural timbers in jetties and wharves, e.g. piles and walings, are almost invariably formed with bolted joints.

Fender piles

Fender piles shall be 300 mm × 300 mm (12 in. × 12 in.) in size and shall

More detailed clauses covering timber piles are

be driven not less than 3 m (10 ft) into the river bed. Twin fender piles shall be notched on adjoining surfaces to receive 150 mm × 100 mm (6 in. × 4 in.) teak shear pins, which shall be driven into position after the twin piles have been securely bolted together. Twin fender piles shall be driven as a single pile, 600 mm × 300 mm (24 in. × 12 in.) in size.

The tops of fender piles shall be cut off at the levels indicated on the Drawings. The fender piles to the wharf shall be tied back to the concrete structure with wrought iron straps and steel bolts. The 300 mm × 300 mm (12 in. × 12 in.) fender piles shall be fitted with cast iron shoes weighing not less than 12·5 kg (28 lb) each and the twin piles with shoes weighing not less than 70 kg (160 lb) each, inclusive of straps.

given in Chapter VII. This clause is concerned with the provision of fender piles to protect the sides of jetties and wharves from the impact of vessels.

Twin piles are also included to extend the scope of the clause. Two 300 mm × 300 mm (12 in. × 12 in.) piles are bolted and pinned together, and driven as a single unit with an enlarged shoe at the base.

Rubbing pieces

Rubbing pieces shall be 300 mm × 150 mm (12 in. × 6 in.) in size and shall fit tightly against the fender piles, to which they shall be fixed vertically with ragged spikes. The top and bottom ends of rubbing pieces shall be bull-nosed.

Rubbing pieces are spiked to the outer face of fender piles to protect the latter from damage. The spiking permits reasonably easy replacement.

Walings and braces

Walings shall be 300 mm × 300 mm (12 in. × 12 in.) in size, except the top waling to the wharf which shall be 300 mm × 150 mm (12 in. × 6 in.). Diagonal braces shall be 300 mm × 150 mm (12 in. × 6 in.) with bolted connections to walings and piles, and splay cut ends.

This clause describes the various horizontal and diagonal members and also the packing and distance pieces, which complete the timber framework to the sides of wharves and jetties.

129

Walings shall be fixed to the concrete structure with wrought iron plates and steel bolts as shown on the Drawings. All packing pieces and distance pieces shall be of Columbian pine creosoted under pressure as previously specified.

Guardrails

Construct guardrail of Columbian pine creosoted under pressure as previously specified. The top of the guardrail shall be 1·25 m (4 ft) above jetty deck level and shall consist of three rails, twice chamfered, 100 mm × 100 mm (4 in. ×4 in.) in size, halved and dowelled to 100 mm × 100 mm (4 in. × 4 in.) posts at 2 m (6 ft) centres. The posts and bottom rail shall be bolted to the jetty members with 15 mm ($\frac{1}{2}$ in.) diameter mild steel bolts as shown on the Drawings.

Constructional features of this type are best described complete in single items. Guardrails are more usually constructed of steel standards supporting tubular steel handrailing (see Chapter VIII for details) but a timber guardrail has been taken here to show the approach.

Jetty decking

The jetty decking shall be constructed of greenheart in 175 mm (7 in.) widths × 50 mm (2 in.) thick, with 15 mm ($\frac{1}{2}$ in.) gaps between the boards. The deck boards shall be spiked with 90 mm ($3\frac{1}{2}$ in.) nails to 150 mm × 75 mm (6 in. × 3 in.) greenheart bearers, spaced at 450 mm (18 in.) centres.

Gaps must be left between the deck boards to permit rain or sea water to pass between them instead of lying on the top surface of the boards and eventually rotting them away. The length of nails or brads is usually taken as the thickness of the boards + 40 mm ($1\frac{1}{2}$ in.).

Tarring

The beds of all intersecting timbers, scarf joints and cut ends of timbers shall have two coats of hot coal tar applied to them before the fastenings

The meeting surfaces of adjoining timbers and cut ends need to be treated with preservatives. Tar must be

130

are secured. The whole of the timber-work on completion shall be cleaned down and painted with two substantial coats of hot coal tar. The coal tar shall comply with the requirements of B.S. 3051: Coal Tar Wood Preservation.

applied hot and should comply with the appropriate British Standard.

Creosoting

All timbers shall be cleaned of mud and dirt and incised to assist penetration of creosote, and all labours shall be executed on the timber as far as practicable prior to creosoting. The timbers shall be creosoted under pressure in accordance with B.S. 913: Wood Preservation by means of Pressure Creosoting, and the creosote shall conform to B.S. 144: Coal Tar Creosote for the Preservation of Timber.

This is an alternative to the application of hot coal tar. An essential difference is that the creosote is applied under pressure, whilst the tar treatment was specified as a brush application.

Bolts, nuts, etc.

All bolts shall have square heads and nuts and clean-cut Whitworth threads. Heads shall be solidly forged with the shanks of the bolts perfectly square to the axis of the bolt, and with the under sides of heads in a true plane to take an even bearing all over the steel washers, on which both the heads and nuts of every bolt shall be bedded. Every head and nut on the water face shall be countersunk and no portion of the shank shall come within 25 mm (1 in.) of the surface of the timber.

Plates shall be of the dimensions shown on the Drawings and shall be evenly bedded on timber or concrete throughout their entire area.

This clause specifies in detail the essential characteristics of the bolts to be used for fixing structural timbers. Note that heads and nuts of bolts must be kept back at least 25 mm (1 in.) from the water face of timbers. It is sometimes specified that the hole shall be filled with hard setting bitumen to render it watertight.

Wrought iron for straps, etc., shall be equal to approved samples and certificates of quality and tests shall be submitted to the Engineer before any iron is used on the job.

Measurement

The measurement of structural timbers shall be taken as the net cubical contents of timber in the finished work actually completed in accordance with the Contract Documents. The prices for timberwork shall include for provision, cutting, shaping and moulding of the timber to the required shapes, sections and dimensions, drilling and countersinking for fixings, forming joints of all kinds, creosoting or tarring as specified, and erecting and fixing the timberwork complete in position. Timber prices shall also include all nails, screws, pins and wedges, but separate items will be taken for bolts, coachscrews, straps, etc.

Structural timbers are measured in cubic metres (previously cubic feet), including all labours and joints in accordance with the Standard Method of Measurement. It is difficult for the Contractor to assess the amount of labour involved in a cubic metre or cubic foot of timber. Bolts and metalwork generally are measured separately.

Equipment

(a) *Lighting installation.* The provisional sum of £2000 (two thousand pounds) is included in the Bill of Quantities for builder's work and attendance in connection with the installation of lighting standards, navigation lights and all necessary wiring. This sum shall be expended in whole or in part as the Engineer may direct, or may be deducted if not required.

A provisional sum may be included for the lighting installation if details of the work have not been prepared at the time of drafting the specification.

(b) *Rubber buffers to fender piles.* The prime cost sum of £22,500 (twenty-two

In some cases rubber buffers are to be supplied by

132

thousand five hundred pounds) is provided in the Bill of Quantities for the supply of the following rubber buffers by a firm to be approved by the Engineer:

No. 130 Solid rubber buffers 250 mm (10 in.) in diameter × 250 mm (10 in.) long with a vulcanised rubber sleeve cast in.

No. 60 Ditto, 375 mm (15 in.) in diameter × 450 mm (18 in.) long, ditto.

The Contractor shall take delivery of the rubber buffers at site, and unload, get in, store and fix in the positions indicated on the Drawings.

(c) *Bollards*. The following prime cost sums are provided in the Bill of Quantities for the supply of coated bollards for the wharf, jetty and dolphins, complete with all necessary holding-down bolts, by a firm to be approved by the Engineer.

P.C. sum of £400 (four hundred pounds) for 8 no. bollards type *X* to wharf.

P.C. sum of £3000 (three thousand pounds) for 12 no. bollards type *Y* to jetty.

P.C. sum of £500 (five hundred pounds) for 2 no. bollards type *Z* to dolphins.

The Contractor shall take delivery of the bollards at site, and shall unload, get in, store prior to fixing, transport, hoist, sling and lower into position, and fix true and level on the wharf, jetty and dolphins in the positions shown on the Drawings and in accordance with the manufacturer's instructions and, after fixing, shall fill the bollards solid with fine graded concrete (class B).

a nominated firm and the main contractor is instructed to fix them in the required positions. It is necessary to provide the Contractor with sufficient information to enable him to price the fixing work. The P.C. sum will be based on a quotation obtained from the supplier.

Further prime cost items are provided to cover the supply of three types of bollard by a nominated firm. The main contractor is required to fix the bollards and fill them with concrete after they have been fixed in the correct positions. In this case the Contractor will be able to obtain sufficient particulars of the bollards from the manufacturer's catalogue.

(d) *Motorised capstans*. The prime cost sum of £23,500 (twenty-three thousand, five hundred pounds) is provided in the Bill of Quantities for 3 no. motorised capstans supplied, delivered and fixed on the wharf by a firm to be approved by the Engineer.

The Contractor shall allow for all necessary attendance upon the Subcontractor and shall assist in unloading materials at site, and shall get in, transport on site, provide free storage accommodation and free use of plant, hoist and lower to required positions, perform all general builder's work, and provide all other facilities required by the specialist workmen fixing the motorised capstans.

With this example the nominated subcontractor both supplies and fixes the equipment. The main contractor has to allow in his price for ancillary functions such as handling materials on site and making storage and plant available.

(e) *Rescue chains*. The following prime cost sums are provided in the Bill of Quantities for rescue chains and fittings, each 2·5 m (8 ft) long, supplied and delivered to site, painted with two coats of bituminous paint, and fixed complete by a firm to be approved by the Engineer:

P.C. sum of £250 (two hundred and fifty pounds) for 5 no. rescue chains to wharf.

P.C. sum of £500 (five hundred pounds) for supply of 10 no. rescue chains to jetty.

The Contractor shall allow for all necessary attendance upon the Subcontractor in a similar manner to that described for motorised capstans.

As with the motorised capstans, the nominated subcontractor is both supplying and fixing the equipment, and the main contractor's responsibilities follow a similar pattern.

(f) *Mooring rings*. The prime cost sum of £200 (two hundred pounds) is provided in the Bill of Quantities for 12 no. mooring rings to the wharf, each consisting of a 100 mm (4 in.) diameter ring

This is another example of a prime cost item covering the supply of equipment by a nominated firm and the fixing of the equipment, in this case

bolt with a 100 mm × 100 mm × 5 mm (4 in. × 4 in. × ¼ in.) flat plate washer and nut, for fixing the ring bolt to a timber waling, and a 225 mm (9 in.) diameter loose ring.

These shall be supplied and delivered to the site, painted with two coats of bituminous paint, by a firm to be approved by the Engineer.

The Contractor shall take delivery of the mooring rings at site, unload, get in, store and subsequently fix to greenheart walings on the wharf in the positions shown on the Drawings.

mooring rings, by the main contractor. The Contractor has to insert a price in the Bill of Quantities to cover all labour associated with handling and fixing of the rings.

SUBSIDIARY CARPENTRY ITEMS

Steps

Creosoted redwood steps, 1·25 m (4 ft) wide overall, shall be constructed where shown on the Drawings with 225 mm × 40 mm (9 in. × 1½ in.) treads housed to 275 mm × 50 mm (11 in. × 2 in.) string boards housed and dowelled to 100 mm × 100 mm (4 in. × 4 in.) posts.

Short lengths of open timber steps are sometimes provided to give access to tanks, filters, etc.

Footbridge

A creosoted redwood footbridge shall be constructed across the brook in the position shown on the Drawings and shall be supported on concrete walls and foundations at each end, of the dimensions indicated on the Drawings.

Two stringers shall be used consisting of two 225 mm × 75 mm (9 in. × 3 in.) timbers 4·5 m (15 ft) long, spiked and bolted together and supported on and securely spiked to 200 mm × 75 mm

A small timber footbridge of stringers, bearer blocks, boards, handrail and upright supports, is described in a single comprehensive item. The dimensions of all the timbers are stated, together with their method of fixing.

(8 in. ×3 in.) hardwood bearer blocks 450 mm (18 in.) long, bolted to concrete walls.

The footwalk shall consist of 150 mm × 50 mm (6 in. × 2 in.) boards 1·25 m (4 ft) long, with 25 mm (1 in.) spaces between them, nailed to stringers, together with 75 mm (3 in.) diameter half-round handrails, each supported on three 75 mm × 75 mm (3 in. × 3 in.) posts fixed to stringers with coach bolts.

Scumboards to sewage works

Scumboards shall be constructed of creosoted redwood 300 mm × 40 mm (12 in. × 1½ in.), formed of two 150 mm × 40 mm (6 in. × 1½ in.) boards, close cramped and connected by 100 mm × 40 mm (4 in. × 1½ in.) cross battens, 300 mm (12 in.) long, at 1 m (3 ft) centres, each bolted on with two 15 mm (½ in.) diameter galvanised mild steel bolts 100 mm (4 in.) long, with nuts and washers.

Scumboards are invariably provided to settling, storm-water and humus tanks on sewage disposal works. Timber is, however, gradually being displaced by asbestos cement and fibre glass.

JOINERY

Quality of timber

The softwood for joinery shall be unsorted joinery quality Scandinavian redwood complying with B.S. 1186, Part I and amendments.

B.S. 1186: Quality of Timber and Workmanship in Joinery, covers such matters as moisture content, straightness of grain, sapwood, checks, splits, shakes and knots. It is accordingly unnecessary to detail the requirements in connection with these defects.

136

Quality of workmanship

The quality of joiner's workmanship shall comply with B.S. 1186.

The thicknesses specified for joiner's wrought timbers are, unless otherwise specified, prior to planing and 3 mm ($\frac{1}{8}$ in.) will be allowed from the thickness stated for each wrought face.

All joinery shall be wrought on all faces and finished off by hand with glass paper, with slightly rounded arrises. All doors and other framed work shall be put together immediately on commencement of the general work, but shall not be glued or wedged up until joinery is prepared in readiness for fixing.

The word 'framed' as applied to woodwork is to be understood as including all the best known methods of joining woodwork together by mortice and tenon, draw-pinning with hardwood pins, or other method. The backs of door frames, skirtings and other similar items of joinery shall be painted with one coat of wood primer before fixing. This priming shall be performed on the site but not before the Engineer has approved the joinery, and the prices for the respective items of joinery shall include for this priming.

Clinker concrete fixing bricks shall be used wherever possible for fixing timber frames, etc.; otherwise they shall be securely plugged to walls or fixed to proper grounds where required.

The prices of all joinery work are to include for nails and screws for fixing, complying with B.S. 1202 and B.S. 1210 respectively.

Any joiner's work which shall split,

B.S. 1186 also describes in detail the method of making the various joints in joinery work and of constructing the moving parts of windows, doors and drawers.

Specification clauses of workmanship should include allowances for planed surfaces, moulded edges, framing, priming non-accessible surfaces, fixing of joinery, and defects.

Joinery specifications vary considerably in practice in their scope and contents, but the attached specification clauses should form a useful guide.

fracture, shrink, part in the joints, or show flaws or other defects due to unsoundness, inadequate seasoning or bad workmanship, shall be removed and replaced with sound material at the Contractor's expense.

Windows

Wood casements shall be Messrs X type J with built up sill of a 65 mm × 40 mm (2½ in. × 1½ in.) weathered and throated member tongued to a 70 mm × 70 mm (2¾ in. × 2¾ in.) member. Each window frame shall be fixed with 4 no. wrought iron cramps, 40 mm × 3 mm × 300 mm girth (1½ in. × ⅛ in. × 12 in. girth), built into the brickwork, and the frames shall be bedded in cement mortar and sealed all round in mastic.

All glass shall be ordinary glazing quality clear sheet glass in accordance with B.S. 952 and be free from waves, specks, disfigurements or blemishes of any kind. Putty for glazing in wood frames shall conform to B.S. 544.

All glass shall be accurately cut and fitted into the rebates and shall be well sprigged, puttied and back puttied, and neatly trimmed off to the depth of the rebate. All rebates shall be primed before glazing. Glass in panes not exceeding 1 m² (8 ft²) shall be 4 mm thick (26 oz/ft²) glass and that in larger panes shall be 6 mm thick (¼ in.) glass.

Windows are frequently mass-produced stock pattern casements in wood or metal, obtained from an approved supplier. It is also useful at this stage to describe the method of fixing and sealing around window frames and to specify the types of glass and putty and the method of glazing. This latter provision obviates the need to include a separate glazing section in the specification.

Doors

All flush doors shall be of the sizes shown on the Drawings and shall be obtained from an approved manufac-

The majority of doors used on civil engineering jobs will be stock pattern doors. Flush

turer. Details of construction shall be submitted to and approved by the Engineer before the order is placed. Plywood facing is to be 4 mm ($\frac{3}{16}$ in.) thick for interior use and 6 mm ($\frac{1}{4}$ in.) for exterior use. Plywood is to comply with B.S. 1455. The interior type shall be of first grade, and the exterior type shall be of weatherproof type, first grade. Unless otherwise specified, the outer plies shall be of alder or birch. Synthetic resin adhesives for plywood shall comply with B.S. 1203, and where required to be waterproof shall be of the AX100 type.

All flush doors, except those described as solid flush doors and those having panels for glazing, shall be constructed in accordance with B.S. 459, Part 2, for 43 mm ($1\frac{5}{8}$ in.) stock flush doors, with a 20 mm ($\frac{3}{4}$ in.) continuous birch fillet tongued and grooved to the core and finished flush with the outer faces of the door.

All ironmongery shall be obtained from a firm nominated or approved by the Engineer. Samples of all ironmongery are to be submitted to the Engineer for his approval before ordering. All ironmongery shall be supplied with screws of the same metal and finish as the article with which they are to be used. Each lock is to be provided with two keys and each master key is to be in quadruplicate.

doors are becoming increasingly popular. Constant references are made to British Standards where appropriate, e.g. B.S. 1455: Plywood manufactured from Tropical Hardwoods; B.S. 1203: Synthetic Resin Adhesives (phenolic and animoplastic) for Plywood; B.S. 459, Part 2: Flush doors. Part 1 of the last standard deals with panelled and glazed doors, Part 3 covers fire-check flush doors and wood and metal frames ($\frac{1}{2}$ hour and 1 hour types) and Part 4 deals with matchboarded doors. Ironmongery is generally identified by reference to numbers in a manufacturer's catalogue.

Miscellaneous joinery work

Shelving. The Contractor shall provide 225 mm × 25 mm (9 in. × 1 in.) plain edge wrought softwood shelving supported on 25 mm × 50 mm (1 in. × 2 in.) chamfered bearers plugged to walls.

Various other joinery items may still remain to be covered, such as shelving, cupboards, skirtings, architraves, linings, pipe casings,

Cupboards. The Contractor shall provide six Messrs X type 1 double floor combined cupboard and drawer units, without cladding to top, back and one side, overall size 1 m × 450 mm × 1 m (3 ft × 1 ft 6 in. × 3 ft), plugged and screwed to brick walls.

etc. Specification clauses covering a few of the more common items are included.

Skirtings. Skirtings shall be of 20 mm × 75 mm ($\frac{3}{4}$ in. × 3 in.) chamfered wrought softwood, nailed to 20 mm × 25 mm ($\frac{3}{4}$ in. × 1 in.) splayed tanalised grounds, plugged and screwed to brickwork.

Painting

Materials. The paint shall be of good quality and shall be obtained from one of the following six manufacturers. All paint shall be delivered to the site in sound and sealed containers, labelled by the manufacturer with the following information:
(1) Type of product;
(2) Brand name, if any;
(3) Use for which it is intended;
(4) Manufacturer's batch number.

The Engineer's representative may take samples from painters' kettles for analysis and test. No thinners or other materials shall be added to the paint without the consent of the Engineer.

Knotting shall comply with B.S. 1336. Stopping for interior woodwork shall be putty complying with B.S. 544. For exterior woodwork it shall be white lead paste with red lead to B.S. 217, type B, and gold size to B.S. 311.

Primers for woodwork shall comply with B.S. 2521, while primers for iron

As there are many hundreds of paint manufacturers in this country producing paints of varying qualities, it is common practice to give a selected list of from six to twelve manufacturers from whom the Contractor can make his choice.

It is usual to specify that all paint must be delivered to the site in the sealed containers of the manufacturer and that samples may be taken from the painters' kettles with the object of preventing the adulteration or thinning of the paint prior to use.

Extensive use has been made of British Standards in specifying painting materials and the Code of Practice as a guide to the means of applying paint.

and steelwork shall be one of the following:

(1) Red lead primer to B.S. 2523 type B;
(2) Calcium plumbate primer to B.S. 3698 type A;
(3) Approved zinc chromate primer. Zinc chromate primer shall also be used with aluminium.

All materials shall be kept in dry stores protected from frost. All painting materials used on a particular surface shall be obtained from the same manufacturer. Colour schemes will be prepared by the Engineer.

Workmanship. All surfaces shall be thoroughly cleaned down and approved prior to the application of paint. Wood surfaces shall be knotted, primed and stopped, as necessary, before the application of undercoat. Paint shall be applied strictly in accordance with the manufacturer's instructions and Code of Practice CP 231. Painted work shall be rubbed down between coats and at least 24 hours shall elapse between the application of succeeding coats. No painting shall be carried out on exterior work in wet or foggy weather or on surfaces which are not entirely dry.

Painting prices shall include for all necessary scaffolding, cradles and plant; painting of brackets and supports to pipes, fittings, etc.; touching up and bringing forward worn and bare patches and areas which have been stopped or filled; and taking off and refixing small items of ironmongery such as door knobs.

It is essential that the paint is applied strictly in accordance with the manufacturer's instructions, that the surfaces are dry and that a sufficient period of time is left between the application of succeeding coats of paint.

Premature failure of paint films frequently arises from unsatisfactory or insufficient preparation work and all surfaces must be properly cleaned down, old paint burnt off or otherwise removed where necessary, knots sealed, priming coat applied and holes and cracks stopped before an undercoat of paint is applied.

Chapter Ten

Specification of Roads and Pavings

This chapter covers a wide range of constructional methods adopted in the provision of roads and footpaths. The modern practice is often to refer to roads as 'pavements', following American terminology.

In drafting specifications for roadworks it is probably advisable to subdivide the specification into the component parts or elements of the work. In some cases these sections need further subdivision into materials and construction or workmanship provisions. Alternatively all materials clauses can be grouped together at the beginning of the specification. This procedure has the great merit of avoiding duplication of items. One suitable approach in the drafting of roadwork specifications follows.

MATERIALS required for roads and footpaths.

ROAD BASES, including formation, hard shoulders and possibly soil stabilisation.

FLEXIBLE ROAD CONSTRUCTION: tarmacadam, bitumen macadam and asphalt.

RIGID ROAD CONSTRUCTION: concrete roads, including expansion and longitudinal joints.

ANCILLARY WORK: kerbs, channels, quadrants and edging.

SURFACE WATER DRAINAGE: road gullies, pipework and manholes.

FOOTPATH CONSTRUCTION: tarmacadam, bitumen macadam, asphalt, in situ concrete and paving slabs.

OTHER ITEMS

Grass verges
Street lighting: possibly covered by prime cost items and road crossings for cables.
Fencing: various types.

Readers are referred to Chapter V for more detailed clauses on concrete work and to Chapter XI for detailed clauses on sewers, drains and manholes. For bridgework, readers will find most of the information they require in the chapters covering excavation, concrete, brickwork, masonry, piling and iron and steelwork.

Typical specification clauses follow, covering a selection of the more commonly used materials and constructional methods. Some guidance on alternative methods, where appropriate, will be found in the explanatory notes.

TYPICAL SPECIFICATION CLAUSES   EXPLANATORY NOTES

ROAD-WORKS: MATERIALS

Hardcore for filling

Hardcore shall be natural broken stone, broken brick or other approved hard material, clean and free from any extraneous matter and graded from 225 mm (9 in.) to 75 mm (3 in.).

The actual material used will be largely determined by the materials available in the particular district.

Ashes

Ashes shall be clean, sharp foundry ashes containing not more than 20 per cent of dust and shall be free from foreign matter. Flue dust and destructor ashes will not be accepted.

Ashes are often used as a base for footpath surfacings and sometimes for concrete roads.

Aggregate for granular bases

Aggregate for granular bases shall be clean, hard crushed rock, limestone, slag or gravel of 50 mm (2 in.) or 40 mm ($1\frac{1}{2}$ in.) nominal size.

See Table 1 of B.S. 63: Single-sized Roadstone and Chippings.

Road tar for surface dressing

Road tar for surface dressing shall comply with B.S. 76, type A, and the equiviscous temperature of the tar shall be in accordance with Road Note No. 1: Recommendations for Tar Surface Dressings.

Note the use of Road Notes and other publications of the Road Research Laboratory as a guide to constructional methods.

Cut-back bitumen for surface dressing

Cut-back bitumen shall be obtained from an approved manufacturer and have a viscosity within the range prescribed by the Engineer.

Typical viscosities are 40–60 secs S.T.V. at 40°C in April, May and September, rising to 80–120 secs S.T.V. at 40°C in June to August.

Aggregate for surface dressing

Aggregate for surface dressing shall consist of hard, clean crushed rock or slag of 10 mm ($\frac{1}{2}$ in.) nominal size and shall comply with the grading requirements of B.S. 63.

Chippings used vary from 3 mm ($\frac{1}{8}$ in.) to 20 mm ($\frac{3}{4}$ in.) in size and may be of granite, basalt, limestone, slag or gravel.

Bitumen road emulsion

Bitumen road emulsion for curing cement stabilised bases or sub-bases shall comply with B.S. 434.

B.S. 434 is entitled Bitumen Road Emulsion (Anionic and Cationic).

Tarmacadam

Tarmacadam for carriageways shall consist of 40 mm ($1\frac{1}{2}$ in.) gauge material

Tarmacadam is normally laid in two courses, but single

for the base course and 10 mm ($\frac{1}{2}$ in.) gauge material for the wearing course. The grading of the 40 mm (1$\frac{1}{2}$ in). material shall be in accordance with B.S. 802, Table 1b. The grading of the 10 mm ($\frac{1}{2}$ in.) material shall also conform to B.S. 802 and shall be medium-textured material complying with Table 3 of the Standard. The aggregate shall be hard, durable and clean crushed limestone to the approval of the Engineer, and the binder shall be of tar.

Tarmacadam for footpaths shall consist of 25 mm (1 in.) gauge material for the base course and 10 mm ($\frac{3}{8}$ in.) gauge material for the wearing course. The grading of the materials, and the binder contents and viscosities shall comply with B.S. 1242.

course construction is sometimes used for surfacing and resurfacing work for light to medium traffic. Similar material of smaller gauge is specified for footpath work, but reference would be made to B.S. 1241 where gravel aggregate is required.

Testing requirements are also covered in the British Standard.

Bitumen macadam

Bitumen macadam for carriageways and footpaths shall be of similar material and gauge to that specified for tarmacadam. The grading of the 40 mm (1$\frac{1}{2}$ in.) material shall be in accordance with Table 2 and the 10 mm ($\frac{1}{2}$ in.) material shall comply with Table 5 of B.S. 1621.

The construction of bitumen macadam roads and paths is similar to those in tarmacadam. B.S. 1621 covers bitumen macadam with crushed rock or slag aggregate, while B.S. 2040 covers that with gravel aggregate.

Rolled asphalt

Rolled asphalt (hot process) base and wearing courses shall comply with B.S. 594 and shall have a natural rock aggregate.

Rolled asphalt is composed of asphaltic cement, fine aggregate, coarse aggregate and a filler.

Cold asphalt

Cold asphalt shall be manufactured in accordance with B.S. 1690 with a

Cold asphalt is used as an impervious wearing course,

crushed natural rock aggregate with a maximum size of 5 mm ($\frac{1}{4}$ in.).

as a carpet coat or for patching or sealing.

Waterproof underlay

The waterproof underlay shall consist of either waterproof paper complying with B.S. 1521, class B, or plastic sheeting to grade 250.

Cement and aggregates are covered in Chapter V.

Fabric reinforcement

Steel fabric reinforcement shall comply with B.S. 4483, reference c636, weighing not less than 5·55 kg/m². The reinforcement shall be made up in mat form.

This type of fabric is made up of 8–10 mm main wires at a pitch of 80–130 mm and 6 mm cross-wires at a pitch of 400 mm with electrically welded intersections.

Dowel bars

Dowel bars shall be formed of mild steel complying with B.S. 785, and shall be 20 mm ($\frac{3}{4}$ in.) in diameter, 600 mm (2 ft) in length, clean, straight and free from all deformations.

Dowel bars are often inserted across transverse joints in concrete road slabs to provide load transmission from one slab to the next.

Joint filler

The joint filler shall be pre-moulded and of a type and manufacture approved by the Engineer. It shall be not less than 10 mm ($\frac{1}{4}$ in.) thick. It shall be 25 mm (1 in.) less in depth than the thickness of the concrete slab, and accurately holed to receive the dowel bars.

Specimens shall be subjected to three applications of load at 20 hour intervals to obtain 50 per cent compression. The specimens shall recover at least 70 per cent of their thickness within 2 hours of the release of the last loading.

The joint filler is usually made of non-extruding fibre strips which are capable of adjusting themselves to the changing widths of expansion joints.

Joint sealer

The joint sealer shall be a rubber-bitumen compound of a proprietary brand, approved by the Engineer, and shall be poured in position in accordance with the manufacturer's instructions.

An alternative method of specifying joint sealers is given in the Department of the Environment Specification for Road and Bridge Works, which refers to the U.S. Federal Specification and cone penetration, flow and bond test requirements.

Forms

Forms shall be of either steel or timber with a depth equal to the thickness of the slab.

Timber forms shall be free from warps and twists, of sufficient thickness to ensure adequate rigidity and with a tamping edge which is true to line and level.

Steel forms shall be of approved section and construction, and shall be perfectly straight or suitably curved, with a broad base and sufficient thickness to withstand, without displacement or distortion, the placing and compaction of the concrete. Steel forms shall be provided with an efficient locking device to ensure continuity of line and level through joints and with steel pins to hold them in position.

It is essential that all forms, whether of timber or steel, should be of adequate strength and properly fixed in the correct positions, to withstand the tamping or vibration of the concrete without movement.

Kerbs

Kerbs shall be of naturally coloured precast concrete and shall comply with B.S. 340. They shall contain a granite aggregate and be hydraulically pressed, with a 125 mm × 250 mm (5 in. × 10 in.)

The majority of road kerbs used today are of precast concrete, the strongest being of granite aggregate, and are hydraulically pressed with

147

half-battered section. Purpose-made radius kerbs shall be used for all radii of 12 m (40 ft) and less with the radius clearly marked on one of the unexposed faces.

patterned faces. There are nine different standard sections of concrete kerb.

Channels

Channels shall be of precast concrete with a granite aggregate complying with B.S. 340, and shall be 250 mm × 125 mm (10 in. × 5 in.) in section.

The same British Standard applies as for kerbs. Channels are now often omitted on grounds of expense.

Quadrants

Quadrants shall be of precast concrete with a granite aggregate complying with B.S. 340. They shall be of 300 mm (12 in.) radius, 250 mm (10 in.) deep with a half-battered curved face.

These are also covered by B.S. 340. They are often provided at each side of footpath crossings on the kerb line: 450 mm (18 in.) radius quadrants are also available.

Edgings

Edgings shall be of precast concrete with a gravel aggregate complying with B.S. 340. They shall be 50 mm × 150 mm (2 in. × 6 in.) in section with a half-rounded top.

These are used at the edges of tarmacadam, bitumen macadam and asphalt paths. Sections of 50 mm × 200 mm (2 in. × 8 in.) and 50 mm × 250 mm (2 in. × 10 in.) are also available.

Concrete flags

Concrete flags shall be naturally coloured precast concrete hydraulically pressed with a granite aggregate and shall comply with B.S. 368. The flags shall be 50 mm (2 in.) thick.

Precast concrete flags are made in four sizes: 900 mm × 600 mm (3 ft × 2 ft), 750 mm × 600 mm (2 ft 6 in. × 2 ft), 600 mm × 600 mm (2 ft × 2 ft), and 600 mm × 450 mm (2 ft × 1 ft 6 in.), and in two thicknesses: 50 mm and 63 mm (2 in. and 2½ in.).

Clayware gully pots

Clayware road gullies shall be 375 mm (15 in.) in diameter × 750 mm (30 in.) deep with a rodding eye in accordance with B.S. 539, Part 1.

Precast concrete gullies should be unreinforced and should comply with B.S. 556.

Gully gratings and frames

Gully gratings and frames shall comply with B.S. 497 and shall be curved bar gully gratings and frames, grade B, with a clear opening of 400 mm × 350 mm (16 in. × 14 in.), and weighing approximately 100 kg (2 cwt).

An alternative is to use kerb type gully covers and frames, grade C, with a clear opening of 475 mm × 375 mm (19 in. × 15 in.) and weight of 100 kg (2 cwt) for use in paths and verges. Grade A heavy duty gully gratings and frames, weighing up to 150 kg (3 cwt), are designed for use in main roads.

Grass seed

Grass seed for use on verges, central reserves and side slopes shall be a tested mix from an approved source, supported by certificates of purity and germination. The mixture shall consist of the following types of grass seed in the proportions given:

Perennial rye grass	60 per cent
Red fescue	20 per cent
Smooth stalked meadow grass	10 per cent
Crested dogstail	10 per cent

Different Engineers specify varying mixes largely resulting from past experience. A high proportion of hardy, quick-growing perennial rye grass is needed in these locations.

Fertiliser

The fertiliser for grass seeded areas shall consist of an approved compound containing:
10 per cent Nitrogen
15 per cent Phosphoric Acid
10 per cent Potash

It is good policy to state the three essential ingredients of the grassland fertiliser.

WORKMANSHIP: ROAD BASES

Surface soil

The Contractor shall strip all top soil from the area of carriageways, footways, verges, cuttings and embankments, and shall stack it where directed for future use.

Surface soil or turf should be stripped from the site of the roadworks. Most of it will be required later for soiling verges and other grassed areas.

Excavation

The Contractor shall excavate in any material down to formation level or to such lower level as is required by the Engineer to reach a sound foundation. The spoil shall be used as filling, if approved by the Engineer, or removed from the site.

There is usually an obligation on the Contractor to provide a tip for the disposal of surplus excavated material.

Preparation of formation

The whole area of carriageways whether on fill or in excavation shall be thoroughly consolidated to the satisfaction of the Engineer. In general a 2·5 tonne (2½ ton) smooth wheeled roller will be adequate for this purpose. Any soft places shall be cut out and replaced with suitable hard, dry filling material and again consolidated.

The finished surface of the formation shall be reasonably smooth and free from loose material and shall be approved by the Engineer. The formation shall have gradients and crossfalls parallel to those of the finished road surface and provision shall be made for the removal of surface water that will collect near the edges of the formation. The surface water shall be directed to

It is essential that the formation or subgrade should be carefully formed and consolidated to the correct levels. The type and weight of roller required will vary with the characteristics of the subgrade. Any soft spots within the carriageway limits must be cut out and replaced with suitable consolidated material and arrangements made for disposing of surface water which would otherwise accumulate on the surface of the formation.

It is good practice to seal the formation with a surface dressing of hot tar or bitumen

150

surface-water gully pots or to temporary ditches with suitable outlets formed clear of the formation, through grips at 3 m (10 ft) intervals. The Contractor shall take care to prevent damage to gully pots or blocking of pots with earth or other debris.

Within 24 hours of the final preparation of the formation, or as soon as practicable thereafter, the Contractor shall surface dress the formation with hot tar, as previously specified, at a temperature within the range 93°–127°C (200°–260°F), evenly applied at a rate of 1·2–1·5 l/m² (3–4 yd²/gal) with a distributor complying with B.S. 1707: Hot Binder Distributors for Road Surface Dressing. Immediately after spraying, the binder shall be covered with 5 mm ($\frac{1}{4}$ in.) gravel chippings at the rate of 160 m²/1000 kg (200 yd²/ton), and be adequately rolled with a 2·5 tonne (2$\frac{1}{2}$ ton) roller.

and chippings of gravel, crushed rock or slag. Some Engineers permit the hand application of binder and chippings.

Filling

No filling material shall be laid until the surface soil has been removed. Filling shall consist of approved material from the excavations or hardcore as specified and shall be placed and consolidated in layers not exceeding 225 mm (9 in.) in thickness of loose material. Consolidation shall be effected by the use of a suitable roller or approved heavy plant and during the process of consolidation the Contractor shall ensure that the material used for filling shall have the correct water content to secure maximum compaction.

The final surface of filling shall be carefully levelled and graded, finishing

The filling material will be determined by a number of factors, including depth of fill, class of road and nature of surplus excavated material. The type of plant required for consolidation purposes will likewise be influenced by the depth and type of fill.

151

parallel to the profile of the road surface.

Sub-base

The carriageway sub-base shall consist of ashes, as specified, laid and rolled to a consolidated thickness of 100 mm (4 in.). The roller shall be the heaviest which will compact the sub-base without unduly disturbing the underlying formation.

The sub-base lies between the formation and the road base in flexible construction. With rigid pavements this layer will constitute the base lying between the formation and the road slab. Thicknesses may range from 100 mm (4 in.) to 250 mm (10 in.), depending on the CBR value and frost susceptibility figure for the subgrade.

Water-bound granular base

The base course shall consist of hard, durable igneous rock with a maximum size of 50 mm (2 in.), laid in two layers each with a consolidated thickness of 100 mm (4 in.).

The aggregate shall be mixed with 2–5 per cent of water depending on the type of aggregate in an approved mixer. Compaction shall be effected with an 8–10 tonne (8–10 ton) smooth-wheeled roller with a minimum of 10 passes.

It is generally considered undesirable to lay water-bound granular bases in layers more than 150 mm (6 in.) thick. Alternative forms of construction include dry-bound granular base, low-binder macadam and tar or bitumen-bound granular base.

Lean concrete base

The coarse aggregate shall be washed gravel complying with B.S. 882, Table 3, in an all-in aggregate, with a maximum size of 20 mm ($\frac{3}{4}$ in.). The mix shall be 1 part of Portland cement to 20 parts of all-in aggregate with a water content of

This is weak concrete with a low water content, consolidated by rollers. This gives a strong but rather expensive form of base, but is used extensively for rigid roads to

152

5-6 per cent by weight, mixed in an approved batch-type mixer.

The lean concrete base shall be 200 mm (8 in.) thick and shall be compacted within 1½ hours of laying with a 400 kg (8 cwt) vibratory roller, followed by an 8-10 tonne (8-10 ton) dead-weight roller. Within 1 hour of compaction the surface shall be sprayed with 55 per cent bitumen emulsion at the rate of 0·75 l/m² (6 yd²/gal). The lean concrete base shall attain a minimum strength of 10 MN/m² (1400 lbf/in.²).

carry heavy traffic flows. Readers requiring more information on the grading of aggregates are referred the Department of the Environment Specification for Road and Bridge Works.

Soil cement

The basic material shall consist of crushed rock graded in accordance with the Department of the Environment specification, mixed with 10 per cent by weight of Portland cement in an approved paddle type mixer. The moisture content shall be determined using the methods outlined in B.S. 1377.

The soil cement shall be laid 150 mm (6 in.) thick and shall be compacted within 2 hours of mixing, starting with a 2-3 tonne (2-3 ton) smooth-wheeled roller and following with an 8-10 tonne (8-10 ton) smooth-wheeled roller.

The soil cement shall be cured by spraying the stabilizing layer with 55 per cent bitumen emulsion at the rate of 0·75 l/m² (6 yd²/gal) as soon as compaction is complete.

A variety of materials can be used for soil cement or cement stabilization, including naturally occurring soil, washed or processed granular material, crushed rock or slag, and pulverised fuel ash or burnt colliery shale. Where suitable soil exists on the site, mix-in-place construction can be adopted, whereby cement is spread ahead of the mixer after the soil has been pulverised. Methods of testing are detailed in B.S. 1924.

ROAD SURFACINGS

Tarmacadam

The carriageway surfacing shall consist of two coats of tarmacadam, machine-laid to a consolidated minimum thickness of 80 mm ($3\frac{1}{4}$ in.), with limestone aggregate and a tar binder. The base course shall consist of 40 mm ($1\frac{1}{2}$ in.) nominal size material as specified, rolled to the required profile with a 10 tonne (10 ton) roller, with a width of roll of not less than 450 mm (18 in.), to a consolidated thickness of 65 mm ($2\frac{1}{2}$ in.), in the manner described in B.S. 802.

The wearing course shall be of medium-textured material as specified to a nominal size of 15 mm ($\frac{1}{2}$ in.). The wearing course shall be rolled to the required profile with a 10 tonne (10 ton) roller to a consolidated thickness of 20 mm ($\frac{3}{4}$ in.). The finished surface of the tarmacadam shall be blinded immediately after consolidation with approved grit of a grading not exceeding 3 mm ($\frac{1}{8}$ in.) to dust. The grit shall have been coated with 2–3 per cent of tar of e.v.t. not exceeding 30°C (86°F).

Where construction joints occur in the surfacing, the edge of the previous work shall be cut back to a vertical face which shall be coated with hot tar before any new work is laid against it. The adjoining edges of manhole covers and similar fittings shall be cleaned and painted with hot tar. The surface across joints shall be tested for truth of level and any irregularities exceeding 5 mm ($\frac{3}{16}$ in.) with a 3 m (10 ft) straight edge shall be rectified immediately.

Tarmacadam surfacing can take various forms, from single-course to two-course work and from open- to close-textured material. The type of aggregate and binder can also be varied.

The courses must be laid to the correct profiles and adequately consolidated, usually with a 10 tonne (10 ton) roller.

B.S. 802 gives details of the grading of the aggregate and the binder content.

Bitumen macadam

The carriageway surfacing shall consist of two coats of open-textured bitumen macadam machine laid to a consolidated minimum thickness of 85 mm (3½ in.), with limestone aggregate and cut-back asphaltic bitumen binder. The base course shall consist of 40 mm (1½ in.) nominal size material as specified rolled to the required profile with a 10 tonne (10 ton) roller, with a width of roll of not less than 450 mm (18 in.), to a consolidated thickness of 65 mm (2½ in.), in the manner described in B.S. 1621.

The wearing course shall be of 15 mm (½ in.) nominal size material complying with Table 5 of B.S. 1621. The wearing course shall be rolled to the required profile with a 10 tonne (10 ton) roller to a consolidated thickness of 25 mm (1 in.). The surface of the newly compacted bitumen macadam shall be blinded with bitumen-coated limestone grit not exceeding 3 mm (⅛ in.) nominal size.

Where construction joints occur in the surfacing, the edge of the previous work shall be cut back to a vertical face which shall be coated with hot bitumen before any new work is laid against it. The adjoining edges of manhole covers and similar fittings shall be cleaned and painted with bitumen.

The accuracy of the finish in the longitudinal direction shall be checked with a 3 m (10 ft) straight edge and the maximum permissible gap between the underside of the straight edge and the bitumen macadam shall be 10 mm (½ in.)

Various constructional methods are available, using single-course or two-course construction, and the bitumen macadam can be open- or close-textured.

A wide range of aggregates can be used, as indicated in B.S. 1621, and the binder can be either cut-back or straight-run bitumen. An alternative material for blinding is fine cold asphalt to B.S. 1690. The total thickness of two-course construction varies from 75 mm (3 in.) to 115 mm (4½ in.) for normal work.

Note the alternative and more comprehensive method of specifying maximum permissible surface irregularities to that adopted for tar macadam.

155

for base course and 5 mm ($\frac{1}{4}$ in.) for wearing course.

Rolled asphalt (hot process)

Rolled asphalt surfacing to carriageways shall consist of two coats of hot rolled asphalt machine-laid to a consolidated minimum thickness of 100 mm (4 in.), all in accordance with the general requirements of B.S. 594. The base course shall have a consolidated thickness of 65 mm ($2\frac{1}{2}$ in.) and shall consist of 70 per cent of an approved crushed rock coarse aggregate and an asphaltic cement of bitumen of 40/60 penetration.

The wearing course shall have a consolidated thickness of 40 mm ($1\frac{1}{2}$ in.) and shall consist of 30 per cent of an approved crushed rock coarse aggregate and an asphaltic cement of equal proportions by weight of bitumen of appropriate penetration and refined Lake asphalt.

The wearing course shall, except under exceptional conditions, be laid within 3 days of the laying of the base course. The base course shall have a clean surface and no traffic, except that required in connection with the wearing course, shall be permitted on it. The surface of the newly laid wearing course shall be blinded with 20 mm ($\frac{3}{4}$ in.) coated chippings rolled into the surface at a uniform rate of between 70 and 90 m²/tonne (90 and 110 yd²/ton), with a roller weighing not less than 5 tonnes (5 tons).

The accuracy of the finish in the longitudinal direction shall be checked with a 3 m (10 ft) straight edge and the maximum permissible gap between the

Most of the essential requirements relating to the manufacture and laying of hot rolled asphalt are detailed in B.S. 594. Some amplification is, however, necessary as outlined in the accompanying specification clauses.

Hot rolled asphalt can be laid as a single course or in two courses and it is advisable to produce a roughened surface by applying pre-coated chippings.

underside of the straight edge and the asphalt wearing course shall be 5 mm ($\frac{3}{16}$ in.).

Cold asphalt

Cold asphalt shall be manufactured and laid in general accordance with B.S. 1690, with a limestone aggregate. Cold fine asphalt shall be machine-laid in a single wearing course with a consolidated thickness of 20 mm ($\frac{3}{4}$ in.) to the required profiles.

Pre-coated bitumen stone chippings (10 mm ($\frac{1}{2}$ in.) single size) shall be evenly distributed over the newly laid asphalt at the rate of 80–130 m^2/tonne (100–160 yd^2/ton). The asphalt and chippings shall then be compacted with a roller weighing not less than 5 tonnes (5 tons). Road channels, 250 mm (10 in.) wide, shall be left clear of chippings.

When laying cold asphalt against existing work, the old material shall be cut back to provide a dense vertical face which shall be painted with hot bitumen.

Fine cold asphalt is used for a variety of purposes, but particularly as carpet coats and for blinding, sealing and patching. Its great advantage is its ease of application and it is now used extensively in the repair and maintenance of roads and paths. It has a life of up to 10 years.

Surface dressing

Hot tar shall be evenly applied to the carriageway surface by means of a distributor complying with B.S. 1707: Binder Distributors for Road Surface Dressing, at the rate of 0·75–1 l/m^2 ($4\frac{1}{2}$–6 yd^2/gal.). The temperature of the tar at the time of application to the road surface shall be appropriate to the equiviscous temperature of the tar being used, but in no case shall it fall below 82°C (180°F).

Immediately after spraying, the tar

The rate of application of binder and chippings varies with the condition of the surface which is being dressed and the type and size of chippings. For initial surface dressing of an open-textured coated macadam wearing course, 5 mm ($\frac{1}{4}$ in.) slag chippings applied at the rate of 110–140 m^2/tonne (140–170 yd^2/ton) and tar at the

157

shall be covered with an even layer of 10 mm ($\frac{1}{2}$ in.) clean slag chippings applied at the rate of 70–80 m²/tonne (90–100 yd²/ton). The chippings shall be rolled with a roller weighing 6–8 tonnes (6–8 tons), rolling being continued until a uniform compact surface has been obtained. When required the Contractor shall mix an approved adhesion agent with the binder prior to spraying.

rate of 0·6–0·7 1/m² ($6\frac{1}{2}$–8 yd²/gal.) would be suitable. Where a hot cut-back bitumen is used as binder, the temperature of application would be 121–177°C and the rate of spread of the binder would be 0·6–0·9 1/m² (5–8 yd²/gal.) depending upon conditions. (See Department of the Environment: Specification for Road and Bridge Works.)

NOTE: Other suitable applications for periodic surface dressing would be:

Type of aggregate	Rates of application	
	Tar	Aggregate
20 mm ($\frac{3}{4}$ in.) slag or crushed rock	1–1·3 1/m² ($3\frac{1}{2}$–4 yd²/gal.)	60–70 m²/tonne (70–80 yd²/ton)
15 mm ($\frac{1}{2}$ in.) gravel	0·9–1 1/m² ($3\frac{1}{2}$–5 yd²/gal.)	70–80 m²/tonne (90–100 yd²/ton)
20 mm ($\frac{3}{4}$ in.) gravel	1·2–1·5 1/m² (3–4 yd²/gal.)	60–70 m²/tonne (70–80 yd²/ton)

CONCRETE CARRIAGEWAY CONSTRUCTION

Waterproof underlay

A waterproof underlay, as previously specified, shall be laid over the base to the concrete carriageway. Laps at joints shall be not less than 150 mm (6 in.) and the underlay shall be free from tears or any other damage when the concrete is deposited upon it.

A waterproof underlay is necessary to prevent loss of water and cement from the concrete slab into the porous base. It must be properly lapped at joints.

Road base levels

Immediately before the underlay is laid the top surface of the road base shall be checked by working a scratch template off the side forms. All irregularities shall be corrected immediately and the surface protected from disturbance in any form.

A final check on road base levels is essential to ensure a constant minimum thickness of concrete road slab.

Forms

Forms shall comply with the requirements previously specified. They shall be set true to line and level on a foundation of sufficient strength to ensure that they will not be disturbed by the placing and compacting of the concrete.

Bent or damaged forms shall not be used. The forms shall be tested for level with a 3 m (10 ft) straight-edge before any concrete is laid and where a variation of more than 3 mm ($\frac{1}{8}$ in.) is found the form shall be taken up and reset. The Contractor shall make good at his own expense any irregularities in the concrete surface resulting from the movement of forms.

Forms shall be set for at least 60 m (200 ft) in advance of the point where concrete is being placed and they shall remain in position for at least 24 hours after concrete has been deposited against them. The Contractor shall set forms to the required radius where the curve has a radius of less than 45 m (150 ft).

A suitable foundation for forms would be a strip of concrete or mortar, 25 mm (1 in.) to 50 mm (2 in.) thick. Forms are kept true to line by the insertion of not less than three pins to each 3 m (10 ft) length of form and by the use of locking joints.

Another way of ensuring that an adequate length of form is fixed in position before concrete is placed is to specify that forms shall be set in position at least 30 hours before concrete is placed against them.

Concrete mix

The concrete used in the road slab shall be class B. When instructed by the Engineer two 150 mm (6 in.) cubes shall

It is normally required that the concrete mix shall be such as to provide concrete of a

159

be made from a representative sample of concrete being delivered at the point of deposit. The average crushing strength of the two cubes shall be not less than 28 MN/m² (4000 lbf/in.²). The cubes shall be made in accordance with the procedure given in Section 6 of CP 114.

If the strength of the cubes is below the specified value, three cores shall be cut from the appropriate section of road slab and tested for crushing strength at an age of approximately three months. If the average strength of the three cores when calculated as equivalent cube strength falls below 31 MN/m² (4500 lbf/in²) the Engineer may order the removal of the concrete at the Contractor's expense.

The water/cement ratio of the concrete shall not exceed 0·55 by weight. The concrete shall be of suitable workability for full compaction and the compacting factor shall be determined in accordance with B.S. 1881: Methods of Testing Concrete.

certain compressive strength at 28 days. Provision must therefore be made for the making and testing of sufficient test cubes of concrete to ensure that the required strength is attained.

Rigid control of the water content of the concrete is also needed to prevent loss of strength.

Specification clauses covering the batching and mixing of concrete are not included here as they have already been given in Chapter V. Class B concrete would probably be a 1:2:4 mix.

Crushing strengths could also be expressed in N/mm²

Placing concrete and reinforcement

The mixed concrete shall be deposited in its final position within 20 minutes of leaving the mixer. It shall be spread uniformly over the road base to a depth of 100 mm (4 in.), with great care being taken to ensure that it is of uniform composition and that no segregation takes place. The concrete shall then be compacted to the 100 mm (4 in.) depth by a vibrated hand tamper.

The steel fabric reinforcement shall be accurately assembled and fitted to-

Concrete road slab thicknesses are affected by the properties of the soil, inclusion or omission of base or steel reinforcement and the intensity of the traffic to be carried.

Concrete road slabs are usually reinforced with a single layer of steel fabric reinforcement positioned about 50 mm (2 in.) down

gether to form a mat on the compacted concrete. The use of small pieces of fabric will not be permitted. All laps shall be one complete mesh and the reinforcement shall terminate 50 mm (2 in.) from the edges and from all joints in the concrete, unless otherwise directed by the Engineer.

After the reinforcement has been placed in position the top 50 mm (2 in.) of concrete shall be spread and tamped to give a total slab thickness of 150 mm (6 in.). Care must be taken to prevent disturbance of the reinforcement and the top layer of concrete shall be spread and tamped within 30 minutes of placing the bottom layer. Spreading and compacting of concrete shall be continuous between joints.

from the top surface of the slab.

Many concrete road slabs are laid with vibrated hand tampers, but with important projects to carry heavy traffic flows concrete spreaders and compacting and finishing machines, of the type described in the Department of the Environment Specification for Road and Bridge Works, may be used.

Compacting concrete

Concrete shall be compacted with a tamper weighing not less than 10 kg/m (7 lb/lin. ft), having a 75 mm (3 in.) tamping edge shod with a steel strip and fitted with an approved type of vibrator. Tampers shall be worked off the side forms and that for the base course shall be recessed at the ends to permit the base concrete to be tamped to a level 50 mm (2 in.) below the finished level of the road.

Tamping shall continue until a uniform closely knit surface is obtained. The surface shall be left slightly serrated except for smooth channels, 250 mm (10 in.) wide, and smooth strips, 65 mm ($2\frac{1}{2}$ in.) wide, adjoining the arrises at joints. Tamping towards a joint shall cease 600 mm (2 ft) from the joint and

It is important that concrete in a road slab should be properly compacted to produce a slightly corrugated road surface to the correct profile.

Strips in the channels need to be finished smooth to offer the least possible resistance to the free flow of water towards gullies, etc.

Tampers need to be of a certain minimum weight per unit length to ensure adequate compaction, and most engineers insist on the use of vibrators.

On large-scale projects the concrete is usually spread by

the remainder shall be tamped working away from the joint.

On completion of tamping, the surface shall be checked with a 3 m (10 ft) straight-edge and any irregularities exceeding 3 mm ($\frac{1}{8}$ in.) shall be corrected immediately.

The channels shall be floated to a smooth finish for a width of 250 mm (10 in.) with a steel float and shall be checked with a straight-edge and spirit level to ensure that there is a true fall to gullies at all points. Any faults disclosed by these checks shall be made good immediately. The concrete around a gully shall be slightly dished towards it and care shall be taken to ensure that there is no obstruction to the free flow of water into the gully. All cast iron frames of manholes, gullies, etc., shall be separated from the concrete by a strip of jointing material not less than 5 mm ($\frac{1}{4}$ in.) thick.

a hopper type of concrete spreader, and a compacting and finishing machine strikes off the concrete to the correct level, compacts the concrete by vibration and/or mechanical tamping, and finishes the concrete by a transverse oscillating screed.

Transverse expansion joints

Each transverse joint shall be an expansion joint located at intervals not exceeding 18 m (60 ft) and at the positions shown on the drawings or where directed by the Engineer. The joints shall be 15 mm ($\frac{1}{2}$ in.) wide, straight and vertical, and extend the full depth of the slab at right angles to the axis of the carriageway. Expansion joints shall be filled with an approved pre-moulded filler of a resilient non-extruding type as specified. This filler shall extend from the underside of the slab to within 25 mm (1 in.) of the top of the slab. The arrises to the adjoining concrete slabs shall be rounded to a radius of 5 mm

The provision of transverse expansion joints is essential to allow the concrete to expand with subsequent rises in temperature.

The width of expansion joints is usually 15 mm ($\frac{1}{2}$ in.) or 20 mm ($\frac{3}{4}$ in.) and the joint filler, of non-extruding material, normally finishes about 15–25 mm ($\frac{1}{2}$–1 in.) from the top of the slab. The joints are subsequently sealed with a rubber/bitumen or other suitable sealing compound.

Dowel bars are often in-

162

($\frac{1}{4}$ in.), by careful use of a suitable edging tool, and there must be absolute truth of level across the joints at all points.

As soon as possible after completion all expansion joints shall be brushed clean, dried and sealed with the specified joint sealer. The sealing compound shall fill the joints flush with the surface of the road, protecting and covering both arrises.

The joints shall be provided with dowel bars, 600 mm (2 ft) in length and of 20 mm ($\frac{3}{4}$ in.) diameter, spaced at mid-depth of the slab parallel to the road surface and to the centre line of the road. The dowel bars shall be spaced at 375 mm (15 in.) centres with both end bars positioned 150 mm (6 in.) from the edges of the slab. The dowel bars shall be bonded to the concrete for one-half of their length and the other half be coated with bitumen. The free ends shall be fitted with a closely fitting closed cap 100 mm (4 in.) long, of non-collapsible metal or waterproof cardboard. A disc of felt or other compressible material, 20 mm ($\frac{3}{4}$ in.) thick, shall be inserted between the end of the dowel bar and the end of the cap to permit horizontal movement of the bar in the sleeve. The bars shall be so fixed that they cannot be displaced during concreting and the joint filler shall have holes drilled in it to accommodate the dowel bars.

corporated to provide load transmission from one slab to the next. Provision must be made for movement of the slabs by securing the dowel bars in one slab but leaving them free to move in the adjoining slab. The size of bars used varies from 20–30 mm ($\frac{3}{4}$-$1\frac{1}{4}$ in.) in diameter.

Longitudinal joints

When road slabs are constructed of two-lane or greater width, the carriageway shall be constructed in longitudinal strips each 3·3 m (11 ft) wide. The longi-

With roads exceeding 5 m (16 ft) in width it is usual to construct the roads in a number of longitudinal strips

tudinal joints shall be tongued and grooved joints of the type shown on the drawings, formed by means of a suitably shaped timber or metal strip attached to the forms.

with suitable joints between them.

Curing concrete

As soon as the concrete surface is complete, it shall be cured by the mechanical application of an approved resinous curing compound at the rate of 0·2–0·3 l/m² (20–25 yd²/gal.). The compound shall be uniformly distributed over the road surface in the form of a fine spray.

The concrete surface shall then be protected against the effects of sun and rain during the setting period by the erection of tents of light movable frames covered with suitable opaque waterproof material, arranged to discharge any water clear of the newly-laid concrete.

The concrete must not be allowed to dry out quickly or it will lose part of its strength and surface cracks will develop, particularly in hot weather or with drying winds. The curing treatment specified is that recommended by the Department of the Environment.

Traffic on finished surface

No vehicular traffic shall be allowed on the finished concrete surface within 20 days of completion where ordinary Portland cement is used or 10 days with rapid-hardening cement, provided that the joints have been permanently sealed in both cases. The Contractor shall take all necessary action to ensure compliance with this clause.

The concrete must be given adequate time to attain sufficient strength before any vehicles are allowed to pass over it. The time period can be reduced by the use of special cements.

ANCILLARY WORK

Kerbs, channels, quadrants and edgings

Precast concrete kerbs, channels and quadrants shall be bedded and jointed in cement mortar on concrete foundations (class C) of the dimensions shown on the Drawings. Immediately after being laid, the kerbs and quadrants shall be haunched on the back face to half their height in concrete, class C, to the dimensions shown on the Drawings. Precast concrete edging shall be supported at the joints by a spadeful of concrete, class D.

The exposed face of kerbs and quadrants shall be not less than 100 mm (4 in.) nor more than 105 mm (4¼ in.) above the channel of the road, except where it is necessary to provide an artificial fall in the channel. The exposed surfaces of kerbs, channels, quadrants and edgings shall conform to the required gradients and curves in the vertical plane and to the required lines and curves in the horizontal plane.

Kerbs, etc. are sometimes laid on a semi-dry concrete foundation, thus dispensing with the need for a mortar bed. With concrete roads it is common practice to lay half-section kerbs on a mortar bed, 15–25 mm (½–1 in.) thick, laid directly on the concrete road slab. To give ample support to the kerbs, it is advisable to provide two 10–15 mm (⅜–½ in.) diameter steel dowel bars, 125–150 mm (5–6 in.) long, to each length of kerb, driven into the concrete while it is still green. The kerbs are then backed up with concrete. Concrete class C would probably be a 1:3:6 mix, and D – 1:12.

Road gullies

Road gullies shall be provided in the positions shown on the Drawings or as directed by the Engineer. Gully pots shall be set truly level on a bed of concrete, class C, 150 mm (6 in.) thick and be surrounded with a minimum of 100 mm (4 in.) of similar concrete. Any backfill around the concrete casing shall be thoroughly consolidated.

The frames to gully gratings shall be set in cement mortar, finishing 5–10 mm

Specification requirements as to the concrete casing of gully pots vary from job to job. Some Engineers specify a 100 mm (4 in.) concrete surround while others require 150 mm (6 in.). In some cases it is specified that the concrete surround is to fill the excavated hole entirely.

The introduction of two or

165

($\frac{1}{4}-\frac{1}{2}$ in.) below the adjoining road surface, on two or three courses of 225 mm (9 in.) brickwork in engineering bricks laid in English bond. Once gullies have been connected to the sewer, the gratings or a temporary cover must be fixed to prevent earth or other material gaining access to the gully or sewer. The Contractor shall clear at his own expense all gullies and/or sewers which become silted or blocked due to non-compliance with this requirement.

three courses of brickwork between the gully pot and the grating gives a measure of flexibility in determining the level at which the gully pot is to be placed.

Electric cable ducts

Excavate for and lay 75 mm (3 in.) diameter asbestos cement cable ducts in the positions shown on the plan or where directed by the Engineer, to a depth of 600 mm (2 ft) below the finished road level. The ducts shall be jointed with detachable sockets and two rubber rings to each joint.

The ends of the ducts shall extend 600 mm (2 ft) beyond the face of the kerb on either side of the carriageway and be sealed with tile or slate stoppers set lightly in cement mortar. The ends of ducts shall be marked with 50 mm × 50 mm (2 in. × 2 in.) timber pegs, 750 mm (2 ft 6 in.) long, surrounded in concrete and extending 150 mm (6 in.) above ground, and shall be inscribed with the letter E and painted a distinctive colour.

It is particularly important to lay service ducts under concrete roads to accommodate future services and so eliminate the need to break up the road slab at a later date. An alternative material is second quality clayware pipes. The ends of the ducts should be clearly marked to avoid much abortive labour in excavation at a later date to expose the ducts.

FOOTPATHS

Tarmacadam footpaths

Tarmacadam footpaths shall be laid on a bed of hardcore with a minimum

The tarmacadam consists of crushed rock or slag with a

166

consolidated thickness of 75 mm (3 in.), blinded with ashes.

The footpath surfacing shall consist of two coats of tarmacadam laid to a minimum consolidated thickness of 50 mm (2 in.) to the required gradients and a crossfall of 1 in 48 towards the kerb. The tarmacadam surfacing shall be laid in accordance with the requirements of B.S. 1242: Tarmacadam 'Tarpaving' for Footpaths, Playgrounds and Similar Works.

The base course shall have a minimum consolidated thickness of 30 mm (1¼ in.) and the wearing course shall be 20 mm (¾ in.) thick, each with grading and binder content requirements complying with Tables 3 and 4 of B.S. 1242 respectively and each rolled with a 4 tonne (4 ton) roller.

Immediately after consolidation the surface of the tarpaving shall be lightly dusted with limestone grit graded 3 mm (⅛ in.) to dust at a rate of not less than 290 m²/tonne (350 yd²/ton). After dusting the surface shall be lightly rolled.

binder of tar or a tar-bitumen mixture.

Tarpaving can be laid in one or two courses, and in the case of a single course the thickness may range from 10 mm (½ in.) to 30 mm (1¼ in.). In two-course work total compacted thicknesses vary from 45 mm (1¾ in.) to 75 mm (3 in.).

The dusting grit may be dry or coated with 2–3 per cent of suitable tar or bitumen, or alternatively fine cold asphalt may be used.

Asphalt footpaths

Asphalt footpaths shall be laid on a 75 mm (3 in.) soil cement base as previously specified. The surface of the soil cement shall be covered with a tack coat of bitumen emulsion applied at the rate of 0·35–0·55 l/m² (10–16 yd²/gal.).

Fine cold asphalt complying with B.S. 1690 shall then be laid with a minimum consolidated thickness of 20 mm (¾ in.), rolled with a 2 tonne (2 ton) roller. After initial consolidation, 15 mm (½ in.) pre-coated limestone chippings shall be

Suitable bases for fine asphalt surfaced footpaths include hardcore, concrete and soil cement. The thickness of the asphalt may vary from 10 mm (⅜ in.) to 25 mm (1 in.). The tack coat may be omitted when the surfacing is being laid on a recently laid base course. The final application of chippings is optional.

rolled into the surface of the asphalt, at the rate of 80–130 m²/1000 kg (100–160 yd²/ton).

Flagged footpaths

Paving flags shall be laid on a 75 mm (3 in.) consolidated base of ashes, rolled with a 2·5 tonne (2½ ton) roller, and a 20 mm (¾ in.) bed of lime mortar.

Precast concrete flags 50 mm (2 in.) thick, as specified, shall be laid to a 150 mm (6 in.) bond with closely butting joints to a crossfall of 1 in 48 towards the kerb, neatly cut to all splays and around all boxes, etc. Standard size flags only shall be used with any 600 mm ×450 mm (2 ft × 1 ft 6 in.) flags used at the back of the path. Where paths are circular in plan the flags are to be laid with cross joints radiating to the point from which the arc is struck. After laying, flags are to be grouted in lime slurry.

Flags can be of precast concrete, stone or reconstructed stone. A good even mortar bed is to be preferred to the five-pat system of bedding. A much sounder job is obtained by grouting the flags on completion.

In situ concrete footpaths

In situ concrete paths are to be laid on a 50 mm (2 in.) consolidated bed of ashes or hoggin. The concrete shall be a 1:2:4 mix with a 20 mm (¾ in.) graded aggregate, finishing 3 in. thick with a lightly tamped surface. Expansion joints shall be provided at 9 m (30 ft) intervals and dummy joints at 2 m (6 ft) intervals. All arrises shall be rounded to a radius of 5 mm (¼ in.).

This is a relatively cheap form of footpath construction. It should be not less than 63 mm (2½ in.) in thickness and should have ample expansion and dummy joints. Various surface finishes are available.

168

OTHER ITEMS

Grass verges

A layer of suitable vegetable soil not less than 150 mm (6 in.) thick shall be spread and levelled over the areas to be grassed to the correct gradients and crossfalls. The vegetable soil shall be free from all rubbish, large stones and other harmful material, shall comply with B.S. 3882 and shall be finished to a fine seed tilth, not less than 50 mm (2 in.) thick.

At an approved time a fertiliser, as specified, shall be distributed at the rate of 0·05 kg/m² (1½ oz/yd²), followed by the specified mixture of grass seed at the rate of 0·06 kg/m² (2 oz/yd²), which shall be lightly raked in and lightly rolled if the surface is dry.

To secure good results at least 150 mm (6 in.) of good vegetable soil raked to a fine tilth is necessary. It is undesirable to lay verges to gradients in excees of 1 in 4 or less than 1·5 m (4 ft 6 in.) wide. Where heavy soil is encountered it is advisable to provide a 100 mm (4 in.) layer of clinker or gravel under the verges.

B.S. 3882 details the texture, soil reaction, stone content and method of testing topsoil. Turf should comply with B.S. 3969.

Chain-link fencing

Chain-link fencing shall be erected to the correct lines with the level of the top of the fencing following roughly the mean level of the ground. The line wires shall be pulled drum tight with proper straining fittings at the straining posts. Each straining post shall be properly strutted in the direction of each line of fence. All straining posts and struts shall be set in concrete, class C, 100 mm (4 in.) thick on all sides for half the depth of the post in the ground.

The chain-link fencing shall comply with B.S. 1722, Part 1 and shall be green plastic coated 1·5 m (5 ft) high × 50 mm (2 in.) mesh × 10½ gauge, tied to four line wires of 9½ gauge with 14 gauge

Chain-link fencing is very popular as it provides a good durable form of boundary. It is made in a variety of heights from 750 mm (2 ft 6 in.) to 2 m (6 ft), with meshes ranging from 30 mm (1¼) to 75 mm (3 in.). Posts can be of steel, concrete or timber.

Cleft chestnut pale fencing to B.S. 1722, Part 4, is suitable for temporary fencing for highway jobs and similar purposes.

Other more permanent types of fencing include close-binding wire ties at 450 mm (18 in.) centres. The posts shall be mild steel angle at 3 m (10 ft) centres of the following dimensions:

Intermediate posts: 44·5 mm × 44·5 mm × 4·7 mm (1$\frac{3}{4}$ in. × 1$\frac{3}{4}$ in. × $\frac{3}{16}$ in.)

Straining posts. 63·5 mm × 63·5 mm × 6·2 mm (2$\frac{1}{2}$ in. × 2$\frac{1}{2}$ in. × $\frac{1}{4}$ in.)

Struts: 44·5 mm × 44·5 mm × 4·7 mm (1$\frac{3}{4}$ in. × 1$\frac{3}{4}$ in. × $\frac{3}{16}$ in.)

All shall be 2·5 m (7 ft 6 in.) long.

boarded fences and wooden post and rail fencing, both of which are covered in B.S. 1722.

Chapter Eleven

Specification of Sewers and Drains

This chapter is primarily concerned with the drafting of specification clauses covering the provision, laying and jointing of sewers and drains in a variety of materials. The specification will also pick up associated operations such as excavation, timbering, backfill, etc., of sewer trenches, concrete protection to pipes and testing pipes. Where large-diameter sewers are to be laid in tunnel, the provision, jointing and grouting of shaft and tunnel segments will probably arise.

A sewer specification would not be complete unless it included clauses covering the construction of manholes, which could be of brick, in situ concrete or precast concrete tube construction. Various items of ancillary equipment must also be covered where applicable, such as benchings, channels, covers, boxsteps, step irons, ladders, safety bars and safety chains.

The chapter will conclude with a few miscellaneous items connected with sewerage schemes and sewage disposal works, such as media for filters, filter distributors and ventilating columns.

As with preceding works sections, a logical sequence of items is advisable and a useful approach follows.

(1) MATERIALS

Various types of pipe and other materials and components such as manhole covers, step irons, bricks, cement, aggregates, steel reinforcement, etc. The last three classes of material will not be covered in the clauses produced in this chapter as they have been covered in earlier chapters.

(2) EXCAVATION

Typical matters to be covered here include trench excavation, trial holes, timbering, supports to crossings, keeping excavations clear of water, excess excavation, backfill and location of pipelines.

(3) PIPELAYING

A list of matters commonly encountered includes loading and unloading pipes, laying pipes, building in pipes, jointing various types of pipe, cutting pipes, keeping pipes free from obstruction, stoppers, concrete protection, thrust blocks and testing pipes.

(4) MANHOLES

These can be subdivided into types and include various ancillary features such as channels, benching, ladders, safety chains and bars, etc. Special provision must be made for any out-of-the-ordinary types of manhole such as back-drop manholes, dual manholes and overflow manholes.

(5) TUNNEL AND SHAFT LININGS

Work under this head can normally be conveniently broken down into three main subdivisions: excavation work; provision of cast iron or precast concrete segments, etc.; and work in assembling, jointing, grouting, etc.

(6) ANCILLARY WORK

Sewer outfalls, ventilating columns, storm overflows and small pumping chambers could come within this category. Many of the detailed specification requirements, such as those relating to excavation, concrete, shuttering, waterproofing, etc., will have already been covered elsewhere. Similarly, the majority of items in sewage works construction have been detailed in earlier chapters under their appropriate works sections.

A selection of typical specification clauses covering the more usual sewerage and drainage items now follows.

TYPICAL SPECIFICATION CLAUSES   EXPLANATORY NOTES

MATERIALS

Glazed vitrified clay pipes

Glazed vitrified clay pipes shall be of British Standard quality complying with B.S. 65 and the necessary test certificates shall be supplied to the Engineer. Clay drain fittings shall comply with B.S. 539, Part 1.

This Standard covers two types of socket: (1) for use with manufacturers' special or flexible joints; (2) grooved or roughened for use with users' jointing materials. Clay pipe diameters range from 75 mm (3 in.) to 900 mm (36 in.).

Concrete pipes

Concrete and reinforced concrete pipes and fittings shall comply with B.S. 556 and shall be provided with an approved type of flexible joint. Straight pipes shall be centrifugally spun. Test certificates shall be supplied to the Engineer.

Normal effective lengths are 1, 1·25, 2 and 2·5 m (3, 4, 6 and 8 ft), and pipe diameters range from 150 mm (6 in.) to 1·25 m (48 in.). Standard tests include hydraulic proof tests, absorption and crushing proof tests.

Spun-iron pipes

Spun-iron pipes shall comply with the requirements of B.S. 1211 for class B pipes and shall have bolted gland joints of approved design. Test certificates shall be supplied to the Engineer.

Pipe-diameters range from 75 mm (3 in.) to 600 mm (24 in.) and the most commonly used lengths are 3·6 m (12 ft) and 5·5 m (18 ft). The tests are hydraulic, mechanical and for straightness.

Cast iron specials

Cast iron specials shall comply with the requirements of B.S. 78 for class B pipes and fittings, and shall have bolted

Class B pipes and fittings are tested to a pressure of 120 m (400 ft) head of water.

gland joints of approved design. Test certificates shall be supplied to the Engineer.

Bolted gland joints give some degree of flexibility, as distinct from caulked lead joints which are entirely rigid.

Pitch-fibre pipes

Pitch-fibre pipes and fittings shall be obtained from an approved manufacturer and shall comply with B.S. 2760. Test certificates shall be supplied to the Engineer.

This type of pipe is being used to an increasing extent for drainage work. Joints are formed of tapered spigots fitting tightly into couplings.

Asbestos cement pipes

Asbestos cement pipes shall comply with B.S. 3656, class 3. Test certificates shall be supplied to the Engineer.

These pipes are made in diameters ranging from 100 mm (4 in.) to 600 mm (24 in.) and in four lengths from 1 m (3 ft 3 in.) to 5 m (16 ft 4¾ in.).

Porous pipes

Porous concrete pipes shall comply with B.S. 1194.

These pipes are used for under-drainage and are made with rebated or O.G. joints in diameters ranging from 75 mm (3 in.) to 900 mm (36 in.).

Bricks

All bricks used in manholes shall be class B engineering bricks complying with B.S. 3921. In addition they shall be hard, sound, square, well burnt, uniform in texture, and regular in shape, with true square arrises and even in size. Care shall be taken in unloading, stacking and handling and no chipped or damaged bricks shall be used. All bricks shall be equal to samples submitted to and approved by the Engineer before any brickwork is commenced.

There are two classes of engineering bricks. Class A bricks have a minimum average compressive strength of $69 \cdot 0$ N/mm^2 (10,000 lbf/in.^{2}), while that for class B bricks is $48 \cdot 5$ N/mm^2 (7000 lbf/in.^{2}). It is advisable to use only the engineering type of brick in manholes owing to the humid and corrosive atmosphere prevailing.

Manhole covers

Cast iron manhole covers and frames shall comply with B.S. 497. Where subject to vehicular traffic they shall be grade A, heavy duty, double triangular type with a 550 mm (22 in.) clear opening and weighing 250 kg (5 cwt), as Table 1 of the Standard. Those not subject to vehicular traffic shall be grade B, medium duty, rectangular solid type and weighing 121 kg (2 cwt, 3 qr, 7 lb), as Table 5.

B.S. 497 covers a wide range of shapes, sizes and weights of manhole covers to suit many different sets of circumstances.

Step irons

Manhole step irons shall be of galvanised malleable cast iron complying with the requirements of B.S. 1247 for 'general purpose pattern' step irons, for building into brickwork or in situ concrete.

This British Standard covers three types of step iron: general purpose pattern; precast concrete manhole pattern; and round bar pattern. Minimum weights of step irons vary from 1·5 kg (3½ lb) to 2·5 kg (5¼ lb) each.

EXCAVATION, ETC.

Trench and manhole excavation

All excavations shall be carried out to the dimensions, levels and gradients shown on the Drawings or as directed by the Engineer, in whatever material may be encountered. Extra payment will be made for excavation in rock as previously defined.

The top-soil shall be excavated separately and be kept separate from the sub-soil, for subsequent replacement on

It is essential that all pipe trenches should be excavated to the required lines and levels. The width of the trenches is usually determined by a number of factors, such as pipe size, depth of trench, type of soil and plant available.

A suitable definition of

175

the ground surface. Paved surfaces disturbed in the course of excavation shall be set aside for future use as directed by the Engineer.

Excavation shall not, in the first instance, proceed closer than 75 mm (3 in.) to formation level. The remaining 75 mm (3 in.) shall be excavated by hand on the same day as the laying of the pipes or concrete bed. The width of pipe trenches shall be adequate to permit the satisfactory laying and jointing of pipes.

rock is given in the excavation specification clause in Chapter IV. Where unstable conditions are encountered at formation level, it is good practice to excavate 75 mm (3 in.) or 100 mm (4 in.) below formation level and fill with concrete.

Trial holes

The Contractor shall excavate all necessary trial holes in advance of pipelaying work and shall backfill them and reinstate and maintain the surfaces. Where prior approval of the Engineer has been obtained to the excavation of trial holes, the Contractor shall receive payment for this work. The Contractor shall, however, at his own expense, take all other reasonable action to determine the position of all underground services likely to affect the pipelaying work.

The excavation of trial holes approved by the Engineer will normally rank for separate payment to the Contractor. Nevertheless, the Contractor can be reasonably expected to obtain all possible information from statutory undertakers about their services with a view to avoiding the need for trial holes as far as possible.

Timbering to excavations

The Contractor shall, at his own expense, provide and fix adequate timbering to support the sides of excavations, and it shall be maintained until the constructional work is sufficiently far advanced to permit its removal. The removal of timbering shall be performed only under the direct supervision of a competent foreman. The Contractor shall be held entirely responsible for any damage or injury resulting from the

The Contractor must provide sufficient timbering to excavations to ensure the safety of the workmen and the permanent works. The cost of timbering is, as a general rule, to be covered in the excavation rates, and would not normally come within the category of 'special items of temporary works'.

inadequacy or premature removal of timbering. Where timbering is left in position on the direction of the Engineer for the permanent support of services or structural work, the Contractor shall be entitled to payment of the additional cost involved.

Supports for existing pipes, etc.

The Contractor shall, at his own expense, adequately timber, shore up, sling or support all pipes or other services in the vicinity of the new pipelines, and shall ensure their continuous effective operation.

Pipes and other services may cross over or under the new pipelines and will then need temporary support to prevent undue strain and possible damage.

Excavations to be kept free of water

The Contractor shall, at his own expense, keep all excavations free from water from any cause by pumping, draining or other means to the satisfaction of the Engineer and for such period as he may require. Any sumps and wells provided for this purpose shall be sited away from the permanent works and shall be subsequently filled with concrete, class D.

The water may arise from a number of sources – watercourses, subsoil water, land springs, existing drains and sewers, etc. The method adopted by the Contractor for the removal of the water will vary with the circumstances.

Excess excavation

Any excess excavation shall be filled with concrete, class D, at the Contractor's expense.

Concrete class D would probably be a 1:10 mix, although some engineers specify 1:3:6.

Backfilling excavations

Excavations for pipe trenches and manholes shall be backfilled in layers not more than 225 mm (9 in.) deep with well compacted and suitable material,

Particular care is needed in the backfilling of pipe trenches, to avoid any possibility of damage being caused

177

with particular care being taken at the sides of pipes. The backfilling material used at the sides of pipes and for a height of 450 mm (18 in.) above them shall be selected material free from large stones and consolidated with narrow wooden rammers. Where mechanical rammers are used to consolidate back-filled material in pipe trenches, the pipes shall be protected by not less than 1 m (3 ft) of hand rammed material. No backfilling shall proceed until the line of sewer has been approved by the Engineer.

to the pipes. The normal pre-cautions include the use of selected material around the pipes and hand ramming in the lower part of the trench.

Location of pipelines

The finally agreed routes and levels of pipelines and the number, depths and locations of manholes may not necessarily coincide precisely with the information given on the Drawings, but will be determined by the Engineer or his representative in the light of information resulting from exploratory excavations or obtained from other sources. The contractor's rates for the permanent work shall be deemed to include for any delay or extra cost which the Contractor may incur as a result of deviations from the planned routes, depths and locations.

Although the Engineer will make every attempt to determine accurately the exact lines and levels of sewers during the pre-contract period, it may, nevertheless subsequently be necessary to make adjustments as further and more positive information concerning the position of other services, etc., comes to light.

PIPELAYING

Loading and unloading pipes

Pipes and specials of all kinds shall be handled with approved lifting tackle when loading or unloading. The Contractor shall not roll pipes down timbers

The Contractor must take care in handling all pipes and specials to prevent damage to them. Pitch-fibre pipes

or inclined ramps without the Engineer's special consent.

Pitch-fibre pipes shall be handled strictly in accordance with the manufacturer's recommendations. Pitch-fibre pipes shall be stacked parallel to one another on level ground, without couplings, with stack heights not exceeding 2 m (6 ft).

need special attention and it is advisable to draw the Contractor's attention to the manufacturer's recommendations.

Laying pipes

Each pipe before laying shall be brushed out and examined, and each cast iron or spun iron pipe shall be tested for soundness by striking with a hammer while the pipe is suspended clear of the ground. When laying pipes, adequate precautions shall be taken to ensure that no bricks, soil or other materials enter the pipes already laid, and a close-fitting stopper shall be placed in the end of the last pipe when work is interrupted.

Pipes shall be laid with the sockets leading uphill and shall rest on even and solid foundations for the full length of the barrel. Socket holes shall be formed in the trench bottom, sufficiently deep to allow the pipe jointer adequate space to work round the pipes, and as short as is practicable to accommodate the socket and permit the joint to be made. In rock or stony ground the pipes shall be evenly laid on a bed of sand, not less than 50 mm (2 in.) thick.

Where pipes are to be laid on a concrete bed, the concrete under and around the pipes shall be laid in a single operation. Precast concrete blocks, one at the back of each socket, or suitably

It is imperative that all sewer pipes should be laid true to line and level, to prevent any risk of blockages when in use. Precautions must be taken to ensure that the barrel of each pipe bears evenly upon a solid foundation, whether of soil or concrete.

Sight rails should be not less than 150 mm (6 in.) deep with their top edges planed true and straight. The centre-line of the pipe should be shown by a vertical line on both front and back faces, and for greater clarity the rail should be painted in contrasting colours.

The boning rods should be accurately made to the various lengths in even feet, with a shoe of sufficient projection to rest on the invert of the last laid pipe. On large diameter sewers laid at flat gradients, the levels of the work should be frequently checked by

wedged bricks for the smaller diameter pipes, shall be used to support the pipes before the concrete is laid. After each block or brick has been properly set and boned in to the correct level, and the pipes laid on them and properly centred and socketed, two hardwood wedges shall be inserted transversely between the body of the pipe and the block, and they shall then be driven together until the pipe is brought to the exact level required.

All pipes shall be of the dimensions, materials and classes shown on the Drawings or as directed by the Engineer and shall be accurately laid to the required lines and gradients. All pipes shall be laid in dead straight lines in both horizontal and vertical planes between manholes. Proper sight rails and boning rods shall be used to ensure that each pipe is laid to the correct levels, and sight rails shall be provided at each change of gradient and not more than 45 m (50 yds) apart.

instrument. See CP 2005: Sewerage.

Building-in of pipes

When pipes are built into brick walls of manholes, semi-circular single ring brick-on-edge arches shall be turned over them.

This provision aims at relieving the pipe of the weight of wall above. An alternative is to use concrete lintels.

Jointing of pipes generally

Experienced pipe jointers only shall be permitted to carry out the work of pipe jointing. Special instructions of the pipe manufacturer, such as with flexibly jointed pipes, shall be followed closely. Jointing of pipes prior to lowering them into a trench will not be permitted

Jointing of pipes is a skilled operation and needs to be performed by experienced workmen. Flexibly jointed pipes are being used to an increasing extent and are particularly well suited

except under special circumstances with the approval of the Engineer.

for use in bad ground conditions, in areas subject to mining subsidence, sea outfalls, etc.

Jointing of clayware and concrete pipes

Pipes to be jointed with yarn and cement mortar shall each be placed well home into the socket of the previously laid pipe. Each shall be jointed with one complete lap of best plain hempen spun yarn dipped in liquid cement mortar and tightly driven home by caulking to occupy not more than one-quarter of the total depth of the socket. The remaining space in the socket shall then be filled with cement mortar (1:2), which shall be trowelled on the face to form a splayed fillet completely covering the exposed end of each socket.

The mortar for jointing shall be of the consistency of putty and shall be well caulked into the socket of the pipe with a wooden caulking tool. The joints shall be clean and smooth on the inside face and the interior surfaces of all pipes shall be kept perfectly clear of all jointing material and other extraneous matter. Joints shall be left undisturbed until set and ready for testing.

Many clayware and concrete pipe sewers are still made with yarn and mortar joints, despite their numerous disadvantages. The joints form the weakest link in the pipeline and thus need to be specified in considerable detail with the object of securing sound, watertight joints. Some engineers specify the angle which the cement fillet shall subtend from the face of the socket (30° or 45°) but the writer prefers the method used in the accompanying clause. A mortar mix of 1:2 produces adequate strength and there is less shrinkage on setting than with a richer mix. There is an increasing tendency to use flexible joints with these two categories of pipe.

Jointing of iron pipes

Cast iron and spun iron pipes shall be jointed with standard plain white hempen spun yarn and good quality soft blue pig lead. At least one complete lap of yarn shall be caulked into the back of the socket to prevent the escape of molten lead and shall not project inside the pipe. The lead shall be run at a single

Standard spigot and socket iron pipes are normally jointed with yarn and molten lead in the manner described in the accompanying specification clauses. Caulking lead or lead-sheathed yarn can be used in place of spun yarn.

pouring to a minimum depth complying with the following schedule.

Diameter of pipe mm (in.)	Minimum depth of lead per joint mm (in.)
Not exceeding 100 (4)	45 (1¾)
125 to 150 (5 to 6)	50 (2)
175 to 200 (7 to 8)	55 (2¼)
225 to 1070 (9 to 42)	60 (2½)
exceeding 1070 (42)	65 (2¾)

The pipe surfaces to be jointed shall be thoroughly cleaned and dried before the lead is poured. The lead after pouring shall project 3 mm (⅛ in.) from the socket to allow for caulking. As soon as the lead is cool it shall be properly caulked with caulking irons and 1·75 kg (4 lb) hammers, finishing with a neat and even surface flush with the face of the socket.

Flanged joints shall be properly made with rubber joint rings complying with B.S. 2494 and mild steel bolts and nuts with the bolts projecting two threads beyond the nuts.

Patent flexible joints shall be made strictly in accordance with the manufacturer's directions.

Fibrous lead may be used as a substitute for run lead, probably in the form of a double collar fibrous-lead joint with spun yarn between the two lead collars, occupying not more than one-third of the total depth of the socket. Fibrous lead is particularly suitable for use under wet conditions and in headings where the use of molten lead might prove dangerous to the workmen.

Jointing of asbestos-cement pipes

Asbestos-cement pipes shall be jointed with approved asbestos-cement screw collars, and specials with suitable detachable joints. All joints shall be made strictly in accordance with the manufacturer's directions. Cut ends of asbestos-cement pipes shall be turned to the correct external diameter for a length of at least 75 mm (3 in.).

Asbestos-cement pipes are used occasionally in drainage work. They have the advantage of easily-formed joints.

182

Jointing of pitch-fibre pipes

Pitch-fibre pipes shall be laid and jointed in accordance with Appendix C of B.S. 2760: Pitch-impregnated Fibre Drain and Sewer Pipes and Fittings. Particular care shall be taken to ensure that all spigots and sockets are absolutely clean and free from grit immediately prior to jointing.

The ends of cut lengths of pitch-fibre pipe shall be finished truly square and spigots shall be properly turned with a special tool supplied by the manufacturer.

Pitch-fibre pipes are being used to an increasing extent particularly on rural drainage schemes. They have many advantages, including long lengths and easily-formed joints, and are particularly well suited for use in bad ground conditions.

Cutting pipes

Pipes shall be cut as necessary to accommodate valves, bends, junctions, etc., and for the proper connection of pipelines. All cuts shall be performed with suitable cutting tools and apparatus.

The main purpose of this clause is to prevent the use of hammers and chisels for cutting pipes.

Junction pipes

Junction pipes shall be inserted in the positions shown on the Drawings or where directed by the Engineer. Any branches which are not immediately connected up shall be closed with joinder caps or clayware discs, set in and filled up to the ends of the sockets with puddled clay and a cement fillet. The position of each junction shall be accurately recorded on a layout plan and in a 'junction book'.

Junction pipes will be needed to pick up present and future connections. The open ends need to be properly sealed and their locations duly recorded.

Concrete protection to pipes

All concrete beds, haunchings and surrounds to pipes shall be of class D

It is often specified that pipes in heading, with 6 m

and extend for 150 mm (6 in.) from the outside surface of the pipe. Concrete beds shall be rectangular in cross section and the concrete shall be carefully packed under and around the pipes, taking care not to displace the pipes in any way. In haunchings the concrete shall be carried up to the horizontal diameter of the pipe and then splayed off tangentially to the top of the pipe. Surrounds shall be carried over the pipes to give 150 mm (6 in.) cover to the barrels at all points.

Concrete beds, haunchings and surrounds to pipes shall be discontinued at flexible joints by the insertion of sheets of approved compressible joint filler.

(20 ft) or more of cover or with less than 1·25 m (4 ft) of cover or 1 m (3 ft) elsewhere shall be surrounded with concrete. Pipes of 450 mm (18 in.) diameter and over (1·25–6 m (4–20 ft) cover) and smaller pipes (4·2–6 m (14–20 ft) cover) are often bedded and haunched in concrete. A common mix of concrete is 1:3:6.

It is now becoming the practice to design the pipes to take the loads rather than to surround them with concrete.

Thrust blocks

All pumping mains and other pipelines that are to be tested at high pressure shall be provided with concrete thrust blocks at all bends, junctions, blank ends, etc., all to the approval of the Engineer. End shuttering may be used but the other faces of thrust blocks shall bear directly against undisturbed ground, and flexible joints must have freedom of movement.

A common mix of concrete for thrust blocks is 1:3:6. It is sometimes specified that the Contractor shall be entirely responsible for any consequences of lack or inadequacy of thrust blocks.

Testing of pipelines generally

Pipelines shall be tested before any concrete haunch or surround is laid or backfilling commenced, except where the latter is necessary for access purposes or for testing at high pressure. The Contractor shall at his own expense uncover any such pipes to remedy leaks or other faults.

As far as possible all pipe joints should be exposed at the time of testing, which must be carried out under suitable conditions and in the presence of the Engineer's representative.

It is good policy to add

All tests shall be carried out in daylight in the presence of the Engineer's representative, using water which is coloured substantially with fluoresceine. Pipelines shall be tested in lengths between manholes or in shorter lengths with the Engineer's approval.

The Contractor shall supply all necessary testing apparatus and water, and shall fill and empty the pipes and dispose of the surplus water. Any pipes showing leaks, sweating or other signs of porosity, shall be condemned and shall be replaced and re-tested at the Contractor's expense.

fluoresceine to the testing water as the distinctive colour will immediately draw attention to any leaks. Special items are sometimes included in bills of quantities to cover the cost of testing work, but in other cases the cost is to be covered by the pipelaying rates.

Testing of non-pressure pipelines

All gravitational pipelines shall be tested with water to a head of not less than 1·5 m (5 ft), measured from the crown of the highest pipe under test. At no point shall the pressure exceed the safe pressure specified for the pipes by the manufacturer.

After allowing a short period for absorption, the vertical pipe at the head of the length under test shall be topped up and the water level observed for not less than 10 minutes. Should the water level drop by more than 25 mm (1 in.), the cause shall be sought and the defect remedied.

After the test, the water shall be released from the stopper while a watch is kept on the vertical pipe, to check that the pipeline and the vertical pipe are unobstructed. No testing shall commence within 48 hours of the making of cement joints.

Where it is difficult to apply a water test, the Contractor may be permitted

The majority of sewers and drains are tested by a water test. A quarter bend is jointed temporarily to the socket of the last pipe laid and vertical pipes or tubes are fitted to the bend to give the desired head. The Code of Practice on Sewerage recommends a minimum head of 600 mm (2 ft) whereas the Code of Practice on Building Drainage advocates a minimum head of 1·5 m (5 ft).

Some engineers advocate the fitting of a vertical tube of not less than half the diameter of the pipes under test. Another alternative specification requirement is that the loss of water under test shall not exceed 1/1000 of the total volume of water in the pipeline.

to apply an air test to gravitational pipe-lines exceeding 375 mm (15 in.) in diameter. The air pressure measured in a manometer tube shall be raised by a hand pump to 100 mm (4 in.) head of water. Should the fall of pressure exceed 25 mm (1 in.) of water in the 5 minutes following the cessation of pumping, the cause shall be sought and the defect remedied.

Testing of pressure pipelines

All pressure pipelines shall be tested by filling with water coloured with fluoresceine and raising the pressure by injecting further water through a manu-ally operated forcepump, fitted with an accurate pressure gauge, until the follow-ing pressures are obtained:

Class D pipes: $1 \cdot 8$ MN/m^2 (260 lbf/in.^{2})
Class C pipes: $1 \cdot 4$ MN/m^2 (200 lbf/in.^{2})
Class B pipes: $0 \cdot 9$ MN/m^2 (130 lbf/in.^{2})

The pump shall then be disconnected from the pipeline and the required pres-sure shall be maintained for 30 minutes.

Pumping mains and pipes in syphons will normally be subjected to this type of test. The pressure must vary with the class of pipe used in the pipeline. The normal period of test is 30 minutes.

Alternatively, the pressures could be expressed in N/mm^2.

Testing of pipelines for obstructions

Pipelines of 525 mm (21 in.) diameter or less shall be tested to ensure that they are free from obstruction after having successfully withstood a water test. A loose spherical or cylindrical plug of the diameter shown in the schedule shall be passed through the whole of the pipe-line. Any obstruction encountered shall be removed or any unevenness of invert shall be made good to the satisfaction of the Engineer.

Tests for straightness and obstruction can be carried out in one of two ways: (1) by rolling a ball or passing a plug through the pipeline, (2) by placing a mirror at one end of the pipeline and a lamp at the other. If the pipe-line is straight, the full circle of light can be observed. The mirror will also show any

Pipe diameter mm (in.)	Plug diameter mm (in.)
100 (4)	95 (3¾)
150 (6)	145 (5¾)
225 (9)	215 (8⅝)
250 (10)	238 (9½)
300 (12)	288 (11½)
375 (15)	363 (14½)
450 (18)	438 (17½)
525 (21)	513 (20½)

obstructions in the pipes. The first method is best suited for testing sewers.

After the test has been successfully performed, a close-fitting stopper shall be placed at the head of the pipeline and shall be kept there until the manhole has been constructed.

MANHOLES

Brick manholes

Brick manholes shall be constructed in accordance with the Drawings and in the positions shown on the layout plan or where directed by the Engineer. Concrete bases shall be of concrete class B, laid on waterproof paper, and of the thickness shown on the Drawings.

The chamber walls shall be constructed in 215 mm (9 in.) brickwork in class B engineering bricks laid in English bond in cement mortar (1:3), with the internal face finished fair and flush pointed.

No bats or broken bricks will be permitted except as closers, and all bricks shall be wetted before use and be laid with the frogs upwards. All bed and vertical joints of brickwork are to be filled solid with mortar and no vertical joint may be flushed up from the top.

Brick manholes are best constructed of engineering bricks finished with a fair face internally. The minimum size of a brick manhole chamber should be 1·25 m × 790 mm (4 ft 1½ in. × 2 ft 7½ in.) and an access shaft should be not less than 790 mm × 675 mm (2 ft 7½ in. × 2 ft 3 in.).

215 mm (9 in.) walls are usually adequate for manholes up to about 3 m (10 ft) deep. For deeper and larger manholes, the walls should be thickened up as the depth increases to take account of the ground and/or water pressure. Channels, benchings, step irons, covers, etc.,

No joints shall exceed 10 mm ($\frac{3}{8}$ in.) in thickness.

Brick courses shall be level and straight with perpends kept in vertical alignment. All brickwork shall be cleaned off after completion and left free from deposits of mortar, etc.

Concrete cover slabs shall be of reinforced concrete, class B, of the dimensions shown on the Drawings.

Precast concrete manholes

Precast concrete manholes shall be constructed on a base of concrete, class B, 150 mm (6 in.) thick laid on waterproof paper, and a base wall of similar class concrete of a thickness equal to the thickness of the chamber rings plus 150 mm (6 in.).

The chamber rings shall be 1·1 m (42 in.) in diameter, jointed in cement mortar (1:2), flush pointed internally and surrounded externally with concrete, class B, to a minimum thickness of 150 mm (6 in.).

Taper rings reducing from 1·1 m (42 in.) to 675 mm (27 in.) diameter shall be of the straight back type and shall be placed so that the bottom of the taper ring is not less than 1·5 m (5 ft) above the top of the manhole benching. Where the chamber is less than 2·75 m (9 ft) deep the taper ting shall be replaced by a precast reinforced concrete slab supplied by the ring manufacturer.

Shaft rings shall be of 675 mm (27 in.) diameter and making-up pieces shall be provided where necessary to bring the precast concrete cover slab to the required level. A brick necking of up to three courses of 215 mm (9 in.) brick-

are dealt with later in the chapter.

Precast concrete manholes can be quickly constructed by unskilled labour and are particularly advantageous in bad ground conditions. The chamber rings are made from 920 mm (36 in.) to 2 m (72 in.) diameter and shaft rings of 675 mm (27 in.) diameter are available in lengths of 150 mm (6 in.) to 1·25 m (4 ft) in multiples of 150 mm (6 in.), all with step irons cast in at the factory.

An alternative to in situ concrete base walls is to use precast concrete invert blocks specially made to suit each individual manhole.

Manholes are occasionally constructed in in situ concrete, where there is considerable repetition work and shuttering can be standardised and re-used. The cover slab should be constructed monolithic with the walls and care must be taken to make the concrete watertight.

work in engineering bricks in cement mortar (1:3) shall be constructed to bring the manhole cover to the correct level.

Benchings and channels

Main channels shall be formed of glazed vitrified clay half-round channels jointed in cement mortar (1:2) to easy curves and true levels. Branch channels shall be curved so as to discharge into the main channel at an angle of not more than 45°. Branch pipes shall be arranged to enter the manhole with their soffits level with the soffit of the main sewer.

Benchings shall be carried up vertically from the channels to the crown level of the largest pipe and then splayed off at a slope of 1 in 12, with the edges rounded to a radius of 20 mm ($\frac{3}{4}$ in.). Benchings shall be constructed of concrete, class B, finished off with a rendering coat of fine granolithic concrete (1:2$\frac{1}{2}$), 25 mm (1 in.) thick, trowelled smooth.

There are two methods in use for the construction of channels: (1) half-round clayware channels, as described; (2) fine concrete rendering in granolithic as described for the benching.

The slope of 1 in 12 to the top of the benching permits workmen to stand on the benching for inspection, rodding and repair work, and also enables deposited solid matter to fall back into the sewer.

Step irons

Step irons as previously specified shall be built into brick manhole walls as the work proceeds, at a spacing of 300 mm (12 in.) centre to centre vertically and 300 mm (12 in.) apart horizontally, in staggered formation.

Another way of describing the positioning of the step irons is 'built into the brickwork every fourth course and staggered 150 mm (6 in.) each side of the centre-line of the shaft walls.'

Boxsteps

Boxsteps shall be approved galvanised cast iron boxsteps, 225 mm ×

These are needed on large diameter sewers to enable

189

150 mm × 160 mm (9 in. × 6 in. × 6½ in.), each weighing 9 kg (20 lb) and built into the concrete benching in the positions shown on the Drawings.

workmen to reach the invert from the top of the benching.

Ladders

Ladders shall be provided to manholes exceeding 4·5 m (15 ft) deep, made of galvanised wrought iron with 65 mm × 25 mm (2½ in. × 1 in.) strings, 300 mm (12 in.) apart in the clear, drilled and fitted with 22 mm (⅞ in.) diameter rungs at 250 mm (10 in.) centres, every fifth rung to have a shoulder formed on the ends. The ends of all rungs shall be neatly riveted over.

The feet of the strings shall be turned outwards for a length of 50 mm (2 in.) and recessed into the concrete benching. At the top of each ladder the strings shall be bent to a radius of 150 mm (6 in.), with a straight length of 215 mm (9 in.) to penetrate the shaft wall, and turned outwards at the back of the wall. Wrought iron stays 65 mm (2½ in.) × 15 mm (½ in.) × 425 mm (17 in.) long shall be set into the manhole walls at intervals not exceeding 1·5 m (5 ft) and fixed to the ladder strings with 20 mm (¾ in.) mild steel bolts.

For manholes exceeding 4·5 m (15 ft) in depth, the Code of Practice on Sewerage recommends the use of galvanised wrought iron ladders in place of step irons. Strings or stringers should be not less than 60 mm × 10 mm (2½ in. × ½ in.) and rungs not less than 20 mm (¾ in.) in diameter. Ladders should be not less than 300 mm (12 in.) between strings with rungs at 250 mm (10 in.) centres.

Details of the method of fixing the ladders must be given.

Where an access shaft exceeds 7·5 m (25 ft) deep, enlarged rest chambers should be provided at 6 m (20 ft) intervals, each with a landing platform incorporating a hinged trap-door under the ladder.

Safety chains

Safety chains for use in manholes shall be galvanised wrought iron close-link, 10 mm (⅜ in.) thick, complying with B.S. 781. The chains shall each have one end fixed to a ring bolt and the other end shall be provided with a suitable hook for securing to a separate 215 mm (9 in.) ring bolt supplied ragged for building into brickwork.

All manholes on sewers of 1 m (3 ft) diameter and over should be provided with safety chains for placing across the mouth of the sewer on the downstream side when men are at work.

Safety bars

Safety bars shall be of galvanised pipe handrail of 40 mm (1½ in.) nominal bore with 215 mm (9 in.) ragged ends for building into brickwork.

Safety bars should be provided at the edges of all benchings, platforms, etc., from which a man might possibly fall into the sewer.

Manhole covers

Manhole covers and frames, as previously specified, shall be securely bedded in cement mortar (1:2). The top surface of the cover and frame shall be flush with the surface of any road, footpath or grass verge in which it is situated, or shall be 75 mm (3 in.) to 150 mm (6 in.) above adjoining ground level in other locations.

Manhole covers need to finish flush with paved surfaces and grassed areas, but in other cases it is advisable to finish several inches above the adjoining ground level for ease of location.

TUNNEL AND SHAFT LININGS

(Taking cast iron segments to shafts and precast concrete segments to tunnels).

Shaft segments

Shaft rings shall be of cast iron of 7 m (23 ft 2½ in.) internal diameter, with each ring 450 mm (1 ft 6 in.) long and consisting of ten ordinary plates, two top plates and one key. The approximate weight of each ring is 4·4 tonnes (4·47 tons). Each plate shall be provided with a 30 mm (1¼ in.) diameter gas plug.

The cast iron rings are made up of a number of segments with bolted and caulked joints between them. Grout holes and plugs are provided in each segment or plate for grouting externally.

Tunnel segments

Tunnel rings shall be of precast reinforced concrete of 4·4 m (14 ft 6 in.) internal diameter, with each ring 600 mm (2 ft) long and consisting of seven ordinary plates, two top plates and one key. Each plate shall be provided with a

Precast concrete rings are of similar construction. See Chapter IV for excavation for tunnel work, tunnel driving and use of compressed air plant and equipment.

191

50 mm (2 in.) diameter grouting hole and plug.

Bolting cast iron segments

The Contractor shall provide grummets of approved design under the steel washers to heads and nuts of bolts, to prevent leakage occurring around the bolts. The tightening of bolts shall be carefully performed to ensure that the grummets are forced well home into the bevel of the bolt-holes and that the washers have an even bearing. Where the rings are erected under compressed air, the compressed air equipment shall be retained in working order until the Engineer is satisfied that the cast iron lining is watertight.

As soon as possible after assembly, the segments are bolted together with mild steel bolts, washers and grummets, to assist in securing a watertight joint.

Caulking longitudinal joints of cast iron segments

Caulking of longitudinal joints shall be carried out as soon as practicable after the cast iron lining is erected. Prior to caulking, the recesses shall be thoroughly cleaned by air jets, water jets, scraping or by a combination of these. Metallic lead shall be used for caulking, of the same width as the width of the recess. Where leakage occurs after caulking, the lead shall be removed and renewed. The remaining parts of the caulking recesses shall be tightly filled and pointed with cement mortar (1:3).

All caulking recesses need to be well cleaned before any lead is applied to the joints. After the strips of lead have been caulked into the joints, it is usual to fill the remaining space in the jointing recesses with cement mortar.

Caulking circumferential joints of cast iron segments

Preliminary jointing shall be carried out with tarred yarn during erection of

The process is similar to that specified for the longithe rings. As soon as practicable after erection, the spun yarn packing shall be removed, the joints shall be cleaned as previously described, and a continuous caulking of metallic lead applied behind the bolts, with lead block joints connecting the circumferential and longitudinal joints. The recesses shall be finished off with cement mortar (1:3) after the lead caulking has been checked for watertightness. The Contractor's prices for caulking circumferential joints shall include for taking out and replacing bolts, providing, cutting out and removing temporary tarred yarn packings, and cleaning out, caulking and pointing the joints.

tudinal joints, except that temporary tarred yarn packings are placed during erection, to be removed subsequently, prior to cleaning out and caulking the joints. It is also necessary to provide a sound connection between the circumferential and longitudinal joints with lead blocks.

When the work is carried out in compressed air, it is usual to caulk the joints temporarily with neat cement prior to grouting.

Grouting outside cast iron lining

The space between the shaft lining and the surrounding ground shall be filled completely with cement grout (1:2), mixed with sufficient water to permit it to be forced by compressed air through holes in the castings. The grouting shall be carried out as soon as practicable after the lining is erected and the joints caulked, and the holes shall be carefully plugged after grouting.

The void outside the tunnel or shaft lining is usually filled with cement grout (1 part cement to 2 parts sand) forced through the grout holes (one to each segment) under pressure.

Jointing precast concrete segments

The segments of precast concrete rings shall be bolted together with mild steel bolts, domed steel washers and approved grummets. In addition, creosoted deal or approved filling compound of 3 mm ($\frac{1}{8}$ in.) compressed thickness shall be placed in the circumferential joints, and approved bitumen filling in the longitudinal joints.

The segments of precast concrete rings are bolted together in a similar manner to cast iron segments, but strips of filling material are also inserted in the joints. The caulking grooves are filled completely with cement mortar.

The caulking groove between the pre-cast concrete segments shall be thoroughly cleaned, wetted and tightly filled with stiff cement mortar (1:3). The finished work shall be watertight on completion and any unsatisfactory caulking or grummetting shall be cut out and replaced at the Contractor's expense.

Grouting outside precast concrete lining

Cement grout, as previously specified, shall be injected under pressure through the holes in the linings, so as to fill completely all the voids between the tunnel lining and the surrounding ground, and the holes shall be carefully plugged after grouting. Grouting shall commence at the invert and proceed upwards, allowing air and moisture to escape through the upper grout holes.

The tunnel linings will be grouted up externally to eliminate any voids outside the linings. Grouting should proceed from the bottom upwards.

Concrete lining to shaft and tunnel rings

A concrete lining of the thickness shown on the Drawings shall be applied to the inner face of shaft and tunnel rings, and finished to produce a dense, hard and smooth surface. The concrete shall consist of one part of sulphate-resisting cement to not more than four parts of total dry aggregate by weight, with a maximum aggregate size of 10 mm ($\frac{3}{8}$ in.) and a water/cement ratio not exceeding 0·50. The concrete shall have a nominal strength of 35 MN/m^2 (5000 lbf/in.^{2}) and a minimum batch cube strength (6 cubes) in accordance with B.S. 1881 of 33 MN/m^2 (4750 lbf/in.^{2}) at 28 days. The compacting fac-

With sewers, culverts and circulating water ducts, it is important that a sound, dense, non-corrosive and smooth interior surface is obtained to the tunnel lining. For this reason it is necessary to specify good quality concrete in every way suited to this purpose.

The reader is referred to the Code of Practice on Sewerage for details of brick-lined concrete sewers.

tor shall not exceed 0·87 and the slump shall not exceed 75 mm (3 in.).

The concrete shall be adequately mechanically vibrated between steel shuttering of approved design. Fabric reinforcement of type B385 to B.S. 4483 shall be provided where shown on the Drawings.

ANCILLARY WORK

Ventilating columns

Ventilating columns shall be of reinforced centrifugally spun hollow concrete, 9 m (30 ft) high overall, 85 mm (3½ in.) diameter nominal bore, to stand 7·5 m (25 ft) above the ground, complete with a 150 mm (6 in.) branch connection, 200 mm × 150 mm (8 in. × 6 in.) galvanised inspection door and frame and galvanised wire balloon.

Alternative materials are asbestos cement and lap-welded steel. The latter is usually coated internally with Dr Angus Smith's solution and painted externally with two coats of bitumastic paint.

Percolating filter distributors

Rotary distributors for percolating filters shall be of 15 m (50 ft) diameter from an approved manufacturer, with four pipe arms, patent air lock water seal, revolving cross head fitted with steel balls and races, galvanised wire ropes and shackles and duck foot bends.

The whole shall be erected, tested and adjusted by the manufacturer.

These are best erected on the site by the manufacturer, as they are specialist items of equipment requiring careful adjustment. Pumps are normally covered by prime cost sums and are installed by the manufacturer.

Media for filters

The media for biological filters shall be air-cooled blast furnace slag or a natural hard stone such as granite, whinstone or basalt, providing that in

Filter media generally consists of the most suitable local material such as clinker, slag, limestone or granite. The

all cases the material shall conform in every respect with the requirements of B.S. 1438, including the sodium sulphate soundness test and all other chemical tests referred to in that Standard and in B.S. 1047, where applicable. Sample loads of media shall be delivered to the site in advance of the main deliveries for approval by the Engineer.

The medium in the bottom 300 mm (12 in.) layer of the filter shall be 100 mm (4 in.) nominal medium kept within the following grading limits:

Passing 150 mm (6 in.) sieve
                    100 per cent
Passing 100 mm (4 in.) sieve
                    95–100 per cent
Passing 75 mm (3 in.) sieve
                    0– 35 per cent
Passing 60 mm (2½ in.) sieve
                    0– 5 per cent

The remainder of the medium shall be 50 mm (2 in.) nominal medium kept within the following grading limits:

Passing 60 mm (2½ in.) sieve
                    100 per cent
Passing 50 mm (2 in.) sieve
                    85–100 per cent
Passing 35 mm (1½ in.) sieve
                    0– 30 per cent
Passing 25 mm (1 in.) sieve
                    0– 5 per cent

The 'index of flakiness' of the 100 mm (4 in.) and 50 mm (2 in.) nominal media shall not exceed 17 per cent and their 'index of elongation' shall not exceed 35 per cent, when determined in the manner described in Appendix D of B.S. 1438.

grading of the media varies from 25 mm (1 in.) to 100 mm (4 in.) in size and the gradings given in the accompanying specification clauses have been extracted from B.S. 1438.

A common depth of filter is 2 m (6 ft). They are often circular in shape and up to 33 m (110 ft) in diameter.

B.S. 1438: Media for Biological Percolating Filters, specifies requirements for durability, grading of sizes, shape and cleanness.

The Water Pollution Research Laboratory has recently experimented satisfactorily with plastic media.

Testing of filter media

The Contractor shall, as and when required by the Engineer or his representative, carry out grading analyses, elongation tests and flakiness tests of the filter media by the methods described in B.S. 1438. The Contractor shall supply all the necessary equipment and keep records of the results of the tests.

The following schedule gives the relevant sizes and gauges of media.

The Contractor is usually required to carry out tests on the media in accordance with B.S. 1438. Sometimes the Engineer specifies the equipment which the Contractor is to supply, such as perforated-plate type sieves, thickness gauges, length gauges and a balance, often weighing up to 45 kg (100 lb) and sensitive to 0·03 kg (1 oz), complete with weights.

Nominal size of medium	Size of pieces in sieved fraction	Thickness gauge	Length gauge
100 mm (4 in.)	Passing 150 mm (6 in.), retained on 100 mm (4 in.)	68 mm (2·70 in.)	202 mm (8·10 in.)
	Passing 100 mm (4 in.), retained on 75 mm (3 in.)	52 mm (2·10 in.)	158 mm (6·30 in.)
50 mm (2 in.)	Passing 75 mm (3 in.), retained on 50 mm (2 in.)	37 mm (1·50 in.)	117 mm (4·50 in.)
	Passing 50 mm (2 in.), retained on 40 mm (1½ in.)	26 mm (1·05 in.)	79 mm (3·15 in.)

Placing media in filters

The Contractor shall take special care to prevent any damage being caused to under-drains or other parts of the filter when placing media in the filter. All filter media shall be screened or hand-forked on site to remove dust and under-sized pieces.

The placing of the media must be done with care to avoid damage being caused to the under-drains, the structure of the filter or the media itself.

Specification clauses cover-

For a depth of 300 mm (12 in.) above the under-drains or tiles, the medium shall be conveyed in wheelbarrows and spread by hand, but above that level the Contractor will be permitted to use narrow gauge light track and small tipping wagons. Alternatively the Engineer may consider the use of other lightweight appliances above the lowest 600 mm (2 ft) of medium. Media shall not be tipped direct from trucks or lorries into the filters.

ing most other constructional work in filters and other sewage works structures will be found in the relevant chapters dealing with earthwork, concrete, brickwork, masonry, etc.

Fibreglass scumboards

Fibreglass scumboards shall consist of a 375 mm × 25 mm (15 in. × 1 in.) angle with a small bulb on the edge of the 375 mm (15 in.) leg, fixed to N 4 aluminium-alloy brackets, 15 mm ($\frac{1}{2}$ in.) thick. The fibreglass shall be 2-ply chopped strand glass-fibre mat, with a minimum density of 0·6 kg/m² (2 oz/ft²) for each ply, with a tissue reinforced resin-rich exterior to all surfaces and a minimum overall resin/glass rate of 3:1. The resin shall be suitable for continuous immersion in sewage with a pH value of between 7·0 and 8·0.

Fibreglass is now beginning to replace timber for scumboards and baffle boxes in settling and other tanks. Typical fibreglass specification requirements are detailed.

Sewage screens

Manually-raked sewage screens shall be 1·65 m (5 ft 6 in.) wide and shall consist of mild steel flat bars 40 mm × 15 mm (1$\frac{1}{2}$ in. × $\frac{1}{2}$ in.) with 20 mm ($\frac{3}{4}$ in.) clear spaces between them, set at an angle of 45° and turned over at the top to facilitate raking, with a transverse bar fixed at each end to maintain the bar spacing. The screens shall rest in 75 mm × 75 mm × 10 mm (3 in. × 3 in. × $\frac{3}{8}$ in.)

This chapter closes with a specification clause covering a small manually-raked sewage screen, of a type frequently encountered in small sewage works, to indicate a suitable method of approach.

steel angles, built into walls, to permit easy withdrawal. The height of the screen measured vertically shall be 2·15 m (7 ft). All parts shall be heavily galvanised after fabrication.

CHAPTER TWELVE

Specification of Pipelines

THERE IS a certain amount of common ground between this chapter and the preceding one dealing with sewers and drains. The present chapter is primarily concerned with the drafting of specification clauses covering the provision and laying of water mains and ancillary work, but most of the clauses will be equally applicable to gas mains and oil pipelines.

Once again the order of presenting the component items of the specification should follow a logical sequence and that given below has much in its favour. It must, however, be emphasised that specifications are peculiar to each job and that while the clauses that follow will form a useful guide in the drafting of pipeline specifications, they cannot possibly cover each and every item that could arise in practice. Most jobs possess some unusual items which need special mention. A schedule of the principal items in a pipeline specification follows.

(1) MATERIALS

Pipes and fittings in a variety of materials; valves for various purposes (sluice valves, air valves, washout valves and hydrants); surface boxes, etc.

(2) PIPELAYING

Excavation, unloading pipes, laying and jointing pipes, watercourse and other crossings, backfilling trenches, surface reinstatement and cutting pipes.

(3) TESTING

Followed by chlorination of mains.

(4) VALVE CHAMBERS

For various purposes, possibly followed by nameplates.

Typical specification clauses covering water mains follow, but in some cases such as excavation of pipe trenches, where similar work has already been covered in Chapter Eleven, readers will be referred to this chapter for detailed specification clauses.

TYPICAL SPECIFICATION CLAUSES   EXPLANATORY NOTES

MATERIALS

Spun iron pipes

Pipes for water mains shall be spun iron pipes complying with B.S. 1211: Centrifugally Cast (Spun) Iron Pressure Pipes for Water, Gas and Sewage, for class B pipes, tested to 120 m (400 ft) head of water and coated both inside and outside with Dr Angus Smith's solution in 3·6 m (12 ft) lengths. Joints shall be bolted gland or similar and approved flexible joints.

Spun iron pipes are made in three classes – B, C and D – tested to 120, 180 and 240 m (400, 600 and 800 ft) head of water respectively. Varying classes of pipe may be required on different parts of a job. Four pipe lengths are available – 3·6 m, 4 m, 4·9 m and 5·5 m (12 ft, 13 ft 1½ in., 16 ft and 18 ft). Flexible joints are now very popular.

Special pipes and castings

Special pipes and castings, including bends, T's and branches, shall comply with B.S. 78: Cast Iron Spigot and Socket Pipes (Vertically Cast) and Spigot and Socket Fittings, class B, tested to 120 m (400 ft) head of water and coated inside and outside with Dr

The specials will need to be of the same class as the adjoining pipes. Flanged spigots and sockets are used for connecting spigot and socket pipes to flanged valves.

A protective coating of Dr

201

Angus Smith's solution. Joints shall be approved flexible joints unless otherwise specified.

T-pieces for air valves shall be spigot and socket with a 150 mm (6 in.) diameter flanged branch drilled to suit the air valve. T-pieces for washouts shall be double socketed with a 100 mm (4 in.) diameter flanged branch, level with the invert and drilled for a sluice valve.

Angus Smith's solution or other approved application is needed to both internal and external surfaces of all pipes and specials to prevent rusting.

Protective coatings on iron pipes, etc.

Any places on coated iron pipes and special castings, either on internal or external surfaces, where the coating is chipped or rusted off, shall be wire-brushed and be given a substantial coating of approved rust-inhibiting composition applied hot.

Provision should be made for re-coating iron pipes and specials where the original coating has become defective.

PVC pipes

PVC pipes shall be unplasticised PVC pipes complying with B.S. 3505, type 1420 and class B with shouldered victaulic joints.

PVC pipes are being used to an increasing extent, particularly in corrosive soils. Readers are referred to Chapter XI for concrete pipes and to B.S. 486 for asbestos-cement pipes.

Steel pipes

Steel pipes shall be lap-welded coated steel pipes complying with B.S. 534, with approved flexible type joints.

Test pressures for steel pipes differ considerably between welded and seamless pipes. Pipes can be covered externally with hessian wrapping or asphalt and may be lined internally with asphalt or concrete.

TYPICAL SPECIFICATION CLAUSES   EXPLANATORY NOTES

Sluice valves

Sluice valves shall comply in all respects with B.S. 1218 as far as applicable and shall be class 1. The direction of closing shall be clockwise.

Main sluice valves shall be 450 mm (18 in.) diameter and shall, where directed, be complete with spur gearing and by-pass with a by-pass valve. Valves shall be double flanged and shall have flanged sockets bolted on: 450 mm (18 in.) diameter sluice valves shall be connected to pipes of larger diameter with cast iron spigot and socket taper pipes, type 1 A.

Sluice valves for pipes smaller than 450 mm (18 in.) diameter shall be double flanged with flanged spigots and sockets bolted on.

Sluice valves for washouts shall be double flanged with a flanged socket bolted to one side only.

All steel parts shall be sherardised after machining or fabrication and all cast iron parts shall be primed with one coat of red lead primer before despatch.

B.S. 1218 covers two classes of sluice valve for 180 and 240 m (600 and 800 ft) head test pressures respectively up to 300 mm (12 in.) diameter. Double socketed valves are sometimes used with spigot and socket pipes; otherwise it is necessary to use adaptors with the double flanged valves. The British Standard details the quality of cast iron, bronze and gunmetal, and the jointing and packing to be used. It also gives the dimensions of spindles, nuts, caps and handwheels. Washouts are sometimes referred to as scours.

Air valves

Double air valves shall be of 150 mm (6 in.) diameter, with screwdown valve combined, rubber balls and brass air vent orifice, supplied by an approved manufacturer.

A common practice is to give a manufacturer's catalogue reference number which provides definite standards of quality, etc.

Hydrants

Hydrants shall be of screwdown pattern to B.S. 750, type 2, with a minimum flow of 2050 lpm (450 gpm) at the hydrant outlet with an inlet pressure of

The requirements for hydrants are usually laid down by the Fire Service, using B.S. 750 as a basis.

203

0·18 MN/m² (25 lbf/in.²). Each hydrant or outlet bend shall be fitted with a suitable boss at the base of the outlet with a self-operating frost valve or drilled plug. The hydrants shall have 60 mm (2½ in.) diameter round thread outlets.

Surface boxes

Sluice valve boxes shall be of cast iron, with a 225 mm (9 in.) square clear opening, 150 mm (6 in.) deep, with a hinged lid, as Messrs X S. 86 or equivalent, with the letters S.V. legibly cast on the top of the cover.

Air valve boxes shall be of cast iron with a 375 m × 325 mm (15 in. × 13 in.) clear opening, 100 mm (4 in.) deep, perforated to facilitate air release and provide ventilation, as Messrs X 14 B or equivalent, with the letters A.V. legibly cast on the top of the hinged cover.

Washout valve boxes shall be as specified for sluice valves, but with the letters W.O. legibly cast on the top of the cover.

Hydrant boxes shall be of cast iron with a 375 mm × 225 mm (15 in. × 9 in.) clear opening, 150 mm (6 in.) deep, and with a hinged cover with the letters F.H. legibly cast on the top of the cover not less than 75 mm (3 in.) high.

Valve boxes and covers vary considerably in their form and dimensions. They may be manufactured from cast iron or semi-steel. Sluice-valve covers may be chained or hinged to the box.

Once again the task of the engineer is simplified by giving a manufacturer's catalogue reference. It is essential that the appropriate letters be cast on the tops of covers for ease of identification.

An alternative with sluice valve boxes is to refer to B.S. 1426 and 3461, type D (225 mm × 225 mm) (9 in. × 9 in. opening) or type E (300 mm × 300 mm) (12 in. × 12 in. opening).

Valve keys, etc.

The Contractor shall supply and deliver to the Engineer's representative, three keys for 100–150 mm (4–6 in.) sluice valves, three keys for 375–450 mm (15–18 in.) valves and three pairs of lifting keys for surface boxes.

It is usual to require the Contractor to supply a number of valve keys and lifting keys for surface boxes.

204

PIPELAYING

Workmanship

The whole of the workmanship necessary for the execution of the works described in the specification shall be of good quality and be undertaken by workmen who are careful, capable and skilled in their various trades or callings according to the class of work upon which they are engaged.

This is a general clause covering all the pipelaying work. The main aim is to ensure that the work is satisfactorily performed by skilled and experienced workmen.

Excavation of pipe trenches

Excavate trenches to the width necessary for the size of pipe to be laid to lines shown on the Drawings or as directed by the Engineer and to the depths required to give cover to the pipes of not less than 1 m (3 ft) from finished levels. Pockets shall be formed for sockets, flanges, valves, etc., so as to give the barrel of each pipe a full bearing throughout its entire length.

Timber or otherwise support the sides of the trenches as may be necessary and keep the trenches free from water.

Detailed pipe trench excavation and trench timbering clauses are given in Chapter XI, to which the reader is referred for more detailed information. The accompanying clause is kept quite brief but embodies all the essential information.

Unloading pipes

All iron pipes and specials shall be unloaded singly from trucks and lorries. Unless hoists are used, pipes shall be unloaded by means of skids and check ropes and no pipe shall be dropped or allowed to roll unchecked. Pipes shall not be permitted to roll together and shall be wedged to prevent further movement.

In laying out pipes on the site, they shall not be allowed to impede traffic or

It is essential that proper precautions should be taken in the unloading of iron pipes and specials to prevent damage to the pipes, etc., or to their external protective coatings. Emphasis has been placed here on cast iron pipes, as many water mains are laid in this material.

The accompanying clause

205

to obstruct paths and accesses to private and other property. Pipes shall not be laid out in beds of ditches and every precaution shall be taken to preserve their cleanliness before laying. Should any pipe become fouled, it shall be washed out and then brushed through with an approved chlorine solution at the Contractor's expense.

also lists the basic requirements to be observed in laying out pipes on the site and the need, with water mains, to cleanse and sterilise thoroughly pipes which become fouled.

Laying and jointing pipes

The pipes shall be laid in straight lines or regular curves so as to avoid any unnecessary resistance to flow. The insides of pipes, etc., are to be cleaned thoroughly before jointing and each iron pipe shall be tested for soundness by being struck with a hammer while the pipe is suspended clear of the ground.

Each pipe shall be carefully lowered onto its prepared bed with slings and tackle. If the prepared bed is damaged or stones are dislodged into the trench, then the pipe shall be raised, and the bed made good and stones removed before the pipe is laid. Any fractured pipes shall be replaced at the Contractor's expense.

Where it is required to shorten a pipe, it shall be cut off cleanly and squarely with an approved pipe cutting machine.

When making screw gland flexible joints, the spigots shall penetrate into the sockets for the required depth, the jointing rings shall be properly fixed and the joints tightened, all in accordance with the manufacturer's instructions, to make thoroughly sound and watertight joints.

In flanged joints, the rubber jointing rings shall be properly fixed and the

Pipes are to be carefully lowered into trenches and laid to the correct lines and levels. The pipes must be kept free from mud, debris or other obstructions during laying and until completion of the Contract. It is often specified that pipes shall be laid singly and that they shall not be jointed before being lowered into the trench.

The accompanying specification clauses cover the making of screw gland flexible joints and flanged joints on iron pipes. Typical requirements for caulked lead joints are detailed in Chapter XI. The jointing of asbestos-cement pipes is also dealt with in the previous chapter. It will probably be necessary to cut pipes to accommodate some of the valves.

bolts pulled up uniformly tight, with two threads projecting beyond the nut and washer.

PVC pipes shall be laid on a 75 mm (3 in.) bed of sand and shall be jointed strictly in accordance with the manufacturer's instructions.

Sluice valves, air valves, washouts and hydrants shall be fixed in the positions shown on the Drawings or where directed by the Engineer and the joints made as specified.

Laying cable and pumping main in same trench

The Contractor shall provide and lay in the same trench as the 525 mm (21 in.) cast iron pumping main, between pumping station A and reservoir B, a rubber-insulated lead-covered single wire armoured four core cable, each conductor consisting of $3/0.7$ mm ($3/0.029$ in.) wires.

After the pipes have been laid and tested and the trench refilled up to 150 mm (6 in.) above the pipes, the cable shall be carefully laid in one side of the trench and covered with 75 mm (3 in.) of selected fine soil and 225 mm $\times$ 115 mm $\times$ 50 mm (9 in. $\times 4\frac{1}{2}$ in. $\times 2$ in.) interlocking cable protection covers as manufactured by Messrs X or other equal and approved. Lengths of cable are to be left at each end as directed by the Engineer for future connection to the transmitting and recording apparatus and for connecting together separate lengths of cable.

The cable shall not come into contact with the main or any other cable or pipe and the Contractor shall make all

The accompanying specification clause covers the precautions to be taken when laying a pumping main and cable in the same trench, as sometimes happens in practice.

It is important to prevent contact between the pipe and cable and to protect the cable with interlocking cable protection covers or other suitable form of covering.

arrangements for carrying the cable under any other cables, pipes, services, etc., which cross the trench above the main.

Laying mains near existing pipes

Where new mains are to be laid alongside, over or under existing sewers, drains, water or gas mains, electric cables, etc., the Contractor shall take care not to disturb the existing pipes and connections to them, and any damage caused shall be made good at the Contractor's expense.

The Contractor shall make arrangements for supporting existing services and for temporarily dealing with the flow in any pipes.

In built-up areas new water mains will often have to be laid in close proximity to other services. In these circumstances it is incumbent upon the Contractor to take all reasonable precautions to avoid damage to the existing services and to undertake any temporary work that may be necessary.

Watercourse and river crossings

Water mains laid under watercourses and rivers shall be surrounded with concrete, class C, and the remainder of the trench shall be completely refilled with clay puddle. The trench through the watercourse or river banks shall also be completely refilled with clay puddle for a length of at least 2 m (6 ft) on each side of the watercourse or river.

The whole of the necessary associated work, including any temporary staging, timbering, cofferdams, piling, pumping, etc., shall be covered in the billed rates and shall be carried out in strict accordance with the requirements of any persons or bodies having jurisdiction over the watercourse or river or its banks, and to the satisfaction of the Engineer.

The Contractor shall be held fully

It is a common requirement that pipes shall be laid under the beds of watercourses and rivers to prevent obstructions to flow in times of flood. It is also advisable to protect the pipe from scour and make the banks watertight on the line of the pipe trench.

A substantial amount of temporary work is often necessary when making watercourse and river crossings.

responsible for any direct or consequential damage which may arise from his operations and shall indemnify the Employer against any claims for damage.

Backfilling and temporary reinstatement of pipe trenches

After the pipes and specials have been inspected, tested and approved by the Engineer or his representative, they shall be properly packed underneath and at the sides up to half pipe height with fine dry material selected from the excavated material, which shall be well rammed with narrow wooden rammers. The filling shall then be carried up to 300 mm (12 in.) above the sockets of pipes with selected material and well rammed, and the remainder of the trench, to within 300 mm (12 in.) of the surface of roads and 150 mm (6 in.) from other surfaces, shall be refilled in well consolidated layers not exceeding 225 mm (9 in.) thick. Mechanical rammers shall not be used within 1 m (3 ft) of pipes.

The tops of trenches shall be filled with material originally taken from the surface and set aside for subsequent reinstatement. The surfaces over trenches shall be maintained relatively level with the adjoining surfaces, with additional material added from time to time as necessary to make up deficiencies due to settlement. The Contractor shall carry out any watering or rolling that may be ordered by the Engineer.

Permanent reinstatement of trench surfaces

The permanent reinstatement of trench surfaces in public highways shall

This is an alternative clause for backfilling of pipe trenches to that given for sewer and drain trenches in Chapter XI. The principal aim in both cases is to ensure the careful ramming of fine material round the pipes to prevent interference with pipes or their joints, and the adequate consolidation of the remainder of the backfilled material in trenches to prevent excessive settlement. Constant attention to temporary reinstatement is needed to avoid the creation of dangerous conditions.

The procedure for the permanent reinstatement of

209

be carried out by the Highway Authority at the expense of the Contractor. The time when the permanent reinstatement is undertaken shall be entirely at the discretion of the Highway Authority.

In the case of trenches in other locations, after the trenches have become thoroughly consolidated to the satisfaction of the Engineer, they shall be permanently reinstated by the Contractor at his own expense, and to the approval of the Engineer and the owner of the land.

The Contractor shall assume full responsibility for the safety of surfaces over all excavations and shall indemnify the Employer against all damage to persons or property by reason of the condition of these surfaces until the expiration of the period of maintenance.

trench surfaces in public highways (roads, paths and verges) varies in different localities. In some cases the Highway Authority carries out the work and in other cases the Contractor is required to do it to the satisfaction of the Highway Authority. The Contractor is usually made responsible for meeting any claims arising from badly reinstated surfaces.

TESTING AND STERILISATION

Testing of pipes

All pipes after being jointed and while uncovered shall be tested with water at a pressure of 120 m (400 ft) head (1·2 MN/m^2) (173 lbf/in.^{2}), in the presence of the Engineer or his representative. The test head shall be calculated from the lowest part of the main under test.

For testing purposes the Contractor shall provide all necessary labour, water, hydraulic pumps with gauges, clips and blank flanges with air outlets and connections for pressure pipes. Adequate support shall be provided to all blanked ends and bends by concreting or other means.

Care shall be taken to release all air

It is essential that the pipeline should be satisfactorily tested with water to the required pressure. Testing periods vary from 30 minutes to 1 hour. The accompanying specification clauses indicate the various precautions which need to be taken.

Some Engineers add a clause to the effect that this testing will not relieve the Contractor from the responsibility of delivering up the whole of the works in a sound, clean and perfect condition, free from leakage or other

210

by slowly filling the mains, with air valves open, before the hydraulic pressure is applied. After the full test pressure has been obtained the pump shall be closed off and the main shall withstand the pressure for one hour. All defective pipes, fittings, joints, etc., shall be made good at the Contractor's expense, after which the testing shall be repeated.

defects under the working pressure.

Chlorination of mains

After the mains have been tested and approved by the Engineer, the water used for testing shall be emptied out and the mains shall then be thoroughly flushed with clean water and again emptied. The mains shall then be slowly filled with clean water containing chlorine solution injected by means of approved portable chlorinating equipment.

After all the mains have been charged with the sterilising solution, they shall remain filled for not less than 12 hours. Thereafter they shall be emptied again and refilled with clean water.

It is essential that all water mains should be sterilised after testing and before they become operative. The accompanying specification clause describes one method of doing this.

VALVE CHAMBERS, ETC.

Sluice valve chambers

Chambers for 18 in. and larger sluice valves with spur gearing and by-pass shall have clear internal dimensions of $1 \cdot 4$ m $\times 1 \cdot 25$ m $\times 1 \cdot 8$ m deep (4 ft 6 in. $\times$ 4 ft $1\frac{1}{2}$ in. $\times 6$ ft 0 in. deep). The base slab shall be constructed of concrete, class B, 150 mm (6 in.) thick, supporting brick chamber walls 215 mm (9 in.) thick in English bond in class B engineering

This clause covers a chamber for a large sluice valve with a by-pass which resembles a manhole in construction.

Smaller sluice valves are specified as being housed in a protecting tube of cast iron, supporting a precast concrete

211

bricks, flush pointed internally, with the top courses corbelled to the reinforced concrete cover slab, 150 mm (6 in.) thick. In the end walls, double brick rings shall be turned over the pipes for the full thickness of the wall. The cover slab shall be of concrete, class B, reinforced with one layer of No. 30 expanded metal, and where the chambers are located in roads, the cover slab shall be further reinforced with two 100 mm × 75 mm × 4·5 kg (4 in. × 3 in. × 10 lb) steel joists.

Holes shall be formed in the cover slab and two surface boxes, as specified, shall be set on the slab and flaunched up with cement mortar, with the top of the covers finishing level with the finished surface of the road, path or verge in which the valve is located, and surrounded with class B concrete, 150 mm (6 in.) wide and 150 mm (6 in.) deep. Sluice valves shall be supported on hardwood wedges on 600 mm × 225 mm × 150 mm thick (2 ft 0 in. × 9 in. × 6 in. thick) precast concrete pads set on the chamber base.

Sluice valves smaller than 450 mm (18 in.) without spur gearing and bypass shall have selected fill well rammed around the valve up to the top flange. An approved cast iron protecting tube shall be supported on the valve flange and a mild steel extension spindle shall be fixed as required. A precast concrete cover slab, 750 mm × 750 mm × 150 mm thick, (2 ft 6 in. × 2 ft 6 in. × 6 in.) suitably holed, shall be set over the protecting tube and shall have a surface box, as specified, set on the slab and flaunched up with cement mortar. Surface boxes shall finish to the required levels and be

cover slab. In both cases surface boxes give access to the valve spindles.

As an alternative to the concrete surround to the surface box, a small area ($\frac{2}{3}$–1$\frac{1}{2}$ m² (yd²)) of granite or whinstone blocks set on rough concrete may be used.

With the smaller sluice valves, a common alternative adopted in practice is to build brick chambers 340 mm × 340 mm (1 ft 1$\frac{1}{2}$ in. × 1 ft 1$\frac{1}{2}$ in.) in the clear with 102 mm (4$\frac{1}{2}$ in.) brick walls set on a 75 mm (3 in.) concrete base. The top courses of the brickwork will be corbelled to receive the surface box, thus eliminating the need for a precast concrete cover slab. Yet another alternative is to use precast chamber rings.

Note: class B concrete would probably be a mix of 1:2:4.

surrounded with concrete as before described.

Air valve chambers

Chambers for double air valves shall have clear internal dimensions of 1·2 m ×900 mm×1·25 m deep (3 ft 9 in. × 3 ft×4 ft deep). The base slab shall be constructed in concrete, class B, 150 mm (6 in.) thick, supporting brick chamber walls 215 mm (9 in.) thick in class B engineering bricks in English bond, mainly flush pointed internally but with the bottom four courses laid dry. Double brick rings shall be turned over pipes.

The cover slab shall be of reinforced concrete, class B, 150 mm (6 in.) thick reinforced with one layer of No. 30 expanded metal. An opening shall be formed in the cover slab and a surface box, as specified, set on it, flaunched up with cement mortar and surrounded with concrete as before described.

Air valve chambers are normally constructed in a similar manner to those for sluice valves, except that it is common practice to leave a few of the bottom courses unjointed to facilitate drainage.

The size of the chamber will vary with the size of the air valve. Precast concrete chamber rings offer a suitable alternative form of construction.

Washouts

Washouts shall be constructed at the low points on the main in the approximate positions shown on the Drawings and where finally determined on the site by the Engineer. 100 mm (4 in.) double flanged sluice valves shall be fixed to double socketed T's having a 100 mm (4 in.) flanged branch level with the invert of the main. 100 mm (4 in.) cast iron pipes shall be laid from the sluice valve to discharge into a watercourse or to some other agreed point.

A chamber shall be constructed around the sluice valve in the manner previously specified.

Washouts or scours are provided at the low points on a water main to permit sediment to be flushed out of the pipes periodically. In some cases hydrants are installed for this purpose. Scour pipes must not be connected to foul sewers.

Hydrant chambers

Hydrant chambers shall be 340 mm × 340 mm (1 ft 1½ in. × 1 ft 1½ in.) internally, built on a concrete base, class B, 75 mm (3 in.) thick. The walls shall be of brick, 102 mm (4½ in.) thick, built of class B engineering bricks, flush pointed internally except for the four bottom courses which shall be laid dry. The top courses of the brickwork shall be corbelled to receive the surface box, as specified, which shall finish level with the adjoining surfaces and be surrounded with concrete as before described.

Hydrant chambers may have to be constructed to the requirements of the Fire Service. The accompanying specification clause is a typical requirement, although it is sometimes specified that weep holes, 25 mm (1 in.) in diameter, are to be provided in each wall as an additional measure to facilitate drainage.

Name plates

Cast iron name plates of approved pattern shall be fixed near all valves. The distance from the plate to the valve shall be accurately measured and shown on the plate to the nearest metre (foot), together with the type of valve. The plates shall be secured to walls of buildings, etc., with strong round-headed spikes driven into wooden plugs, or shall be bolted to suitable reinforced concrete posts 1·25 m (4 ft) long and driven 450 mm (18 in.) into the ground.

All valve name plates shall be painted with one coat of red lead primer before fixing, and after fixing with two coats of good quality lead paint with letters and figures picked out in white on a blue background. Hydrant plates will be supplied to the Contractor by the Fire Service.

All valves need to be clearly identifiable by the provision of prominent name plates or markers on the highway boundary opposite the valve. It is customary to use letters to represent the different types of valve, such as S.V. for sluice valve, A.V. for air valve, W.O. for washout and H for hydrant. Hydrant plates are normally yellow in colour, and water fittings are sometimes indicated by white letters and figures on a black background or black letters, etc., on a white background.

A common practice nowadays is to use loose figures for valve sizes and distances which fit into slots in the plates, all enamelled to avoid the need for painting.

Chapter Thirteen

Specification of Railway Trackwork

THE specifying of railway trackwork can be conveniently broken down into three main processes: (1) preliminary work, (2) track materials and (3) laying the trackwork or permanent way.

The preliminary work consists of site preparation, excavation and fill, drainage, compaction and grading. The materials consist of ballast, sleepers of timber, concrete or steel; rails (bull-head or flat-bottom); chairs; fishplates and ancillary fixing items. More complicated items of equipment such as points and crossings may be specified by referring to specific products of a certain manufacture.

The clauses covering the laying of the permanent way must include reference to all the labours involved in laying the track to the required lines, levels and curves. These clauses are usually followed by information on measurement aspects to simplify the subsequent process of measurement and valuation of the work, and to draw the Contractor's attention to all the matters for which he is to make allowance in his prices.

Typical railway trackwork specification clauses follow. It is hoped that these will form a useful guide in the drafting of specification clauses for this class of work.

TYPICAL SPECIFICATION CLAUSES   EXPLANATORY NOTES

PRELIMINARY WORK

Site preparation

The whole of the area of the site as shown on the Drawings shall be cleared of all obstructions, including trees and

The whole of the site, including the area occupied by slopes to cuttings and

215

undergrowth with their roots, rubbish, etc. The resultant debris shall be burnt or removed from the site and the whole of the area left to the satisfaction of the Engineer.

embankments, needs to be cleared of obstructions. In particular the roots of trees and undergrowth must be removed.

Excavation

Topsoil shall be stripped and stacked for subsequent re-use on the slopes to cuttings and embankments.

Excavation shall be performed to the dimensions shown on the Drawings or as directed by the Engineer or his representative on the site. Suitable excavated material shall be used as fill in embankments and all surplus removed from the site.

The Contractor shall not excavate below formation level, except to cut out soft spots. The soft spots shall be refilled with selected and approved material.

The reader is referred to Chapter IV for more detailed specification clauses covering excavation. The Engineer usually aims at equalising the amounts of cut and fill as far as possible, although this is more difficult in railway trackwork than with roads, due to the flatter gradients and slower changes of grade involved.

Surface-water drainage

The Contractor shall keep the formation free from water by pumping, provision of sumps and drainage channels, etc., all to the approval of the Engineer. Open trenches shall be cut and porous drains laid in the positions and to the details shown on the Drawings.

The formation must be kept clear of water to prevent softening and damage. Open trenches or land drains will be needed on each side of the track in cuttings to intercept and take away the water running down the slopes.

Compaction of formation, etc.

The formation of the track and the fill to embankments, deposited in 225 mm (9 in.) layers, shall be adequately consolidated with a smooth-wheeled roller weighing 8–10 tonnes (8–10 tons).

The formation and layers of filled material in embankments must be adequately compacted. The number of passes of the roller will depend on the weight of roller and type of fill.

Grading of formation and slopes

The formation and slopes to cuttings and embankments shall be accurately graded to the required lines and gradients. Sloping surfaces shall be covered with 150 mm (6 in.) of suitable topsoil and seeded, as previously specified, or covered with a 100 mm (4 in.) layer of approved ballast.

Culverts and pipes under the track shall be completed before the final preparation of the formation is carried out.

The treatment of the sloping surfaces to cuttings and embankments may take one of several forms:

(1) left as excavated or filled;
(2) soiled and seeded;
(3) covered with ballast or other suitable material.

Intercepting trenches or land drains will probably be needed at the tops of cuttings and bottoms of embankments.

TRACK MATERIALS

Ballast

Ballast for permanent way construction shall consist of clean, hard crushed stone or other suitable material approved by the Engineer. Bottom ballast shall be evenly graded from 200 mm (8 in.) to 100 mm (4 in.) and top ballast from 60 mm ($2\frac{1}{2}$ in.) to 25 mm (1 in.). Stones used as ballast shall be roughly cubical in shape and the use of flat stones will not be permitted.

Hard, clean stones are normally required for ballast, although slag and clinker are occasionally used. It is desirable to select a material which will drain satisfactorily and will not break down into dust. The life of sleepers is affected by the quality of the ballast.

Timber sleepers

Timber sleepers shall be of creosoted Douglas fir, Scots pine or other approved timber. They shall be straight, sound, square cut and free from injuries, waney edges, shakes, large and dead knots, decay, insect attack and other

Sleepers may be in timber, steel or concrete. There are British Standards covering the latter two categories but there is not one for timber sleepers. Nevertheless, steps

217

serious defects, and shall contain not more than 15 per cent sapwood taken as an average of both end sections.

The moisture content of timber sleepers shall not exceed 22 per cent of the dry weight at time of use. The timber shall, unless otherwise specified, comply with B.S. Code of Practice 112 (Table 1, Group 1) and the measurable characteristics and moisture content shall be assessed in accordance with B.S. 1860: Structural Timber: Measurement of Characteristics affecting Strength.

Timber sleepers shall be not less than 2·6 m (8 ft 6 in.) long and 250 mm × 125 mm (10 in. × 5 in.) in section. Unless suitably finished the top faces of sleepers shall be dressed under each rail for a width of at least 225 mm (9 in.), where the rails rest directly on the sleepers, and for a width suitable for chairs or bearing plates where these are used. Special creosoted timbers of larger section and greater length shall be used at points and crossings, as directed by the Engineer, and these shall be dressed or finished as specified for sleepers.

All timber sleepers shall be creosoted under pressure in accordance with B.S. 913: Wood Preservation by means of Pressure Creosoting, and the creosote shall comply with B.S. 144: Coal Tar Creosote for the Preservation of Timber.

Holes shall be bored in sleepers and crossing timbers to receive coachscrews, bolts and spikes and the holes shall be 5 mm ($\frac{1}{4}$ in.) less in diameter than the coachscrews, etc. The bolts and coachscrews shall be screwed up tight and the spikes hammered home until a firm bearing on the rails or chairs is obtained, all to the approval of the Engineer.

have been taken in the accompanying specification clauses to make the maximum use of all relevant British Standards, for the sake of uniformity and to ensure the use of good quality materials.

Sound and suitable timber, creosoted under pressure, is an essential requirement, with the object of securing sleepers with a minimum life of from 10 to 20 years, depending on the volume of traffic carried.

Some engineers specify joint sleepers of 300 mm × 125 mm (12 in. × 5 in.) cross section, whereas others require two sleepers of normal section to be positioned close to the joint.

The requirements relating to the characteristics of the timber vary appreciably. For instance, some engineers permit up to 25 per cent sapwood and wanes on each of the two edges of the wide face.

Steel sleepers

Steel sleepers for flat-bottom rails shall comply with B.S. 500, and be of inverted trough form, 280 mm (11¼ in.) wide × 85 mm (3½ in.) deep overall and not less than 8 mm (5⁄16 in.) thick. The chairs shall be an integral part of the sleeper or shall be held in position by suitable metal clips welded to the sleeper. The weight of steel sleepers shall be not less than 25 kg/m (140 lb per length of 8 lin. ft).

B.S. 500 gives details of test samples, templates and gauges, methods of manufacture, cleaning and dipping, and inspection during manufacture. It does not however specify any standard dimensions.

Concrete sleepers

Concrete sleepers shall comply with the details shown on the drawings.

The British Standard covering this type of sleeper has been withdrawn.

Rails

Rails shall comply with the requirements of B.S. 9 for bull-head rails and B.S. 11 for flat-bottom rails. Rails shall weigh 47 kg/m (96 lb/lin. yd) and shall be supplied in 18 m (60 ft) lengths and laid to a gauge of 1·44 m (4 ft 8½ in.).

These Standards cover the quality of material, chemical composition and mechanical properties, conditions of finished rails, locations of holes, and dimensions, shapes and weights. Bull head rails range from 30 to 50 kg/m (60 to 100 lb/yd) and flat bottom rails from 12 to 55 kg/m (25 to 110 lb/yd).

Fishplates

Fishplates shall be of mild steel of the shallow type to suit the section of the rails and of sufficient length to take four bolts, all in accordance with B.S. 47:

Two fishplates are required to each joint between lengths of rail, made up of one plate on each side of the rail. B.S.

Steel Fishplates for Bull head and Flat bottom Railway Rails. They shall weigh not less than 14 kg (32 lb) per pair.

47 covers quality, tests, punching of holes, dimensions and weights.

Rail fixings

All rail fixings shall be of mild steel and shall be of the types and dimensions shown on the Drawings or as directed by the Engineer. Bolts and nuts, coach-screws and spikes, etc., shall comply with B.S. 64: Steel Fishbolts and Nuts for Railway Rails, where applicable, shall be forged from the solid and cleanly cut and, where appropriate, have Whitworth standard threads of uniform pitch. All nuts shall accurately fit the threads of the bolts and shall be hand-tight. The bolts shall be of sufficient length to project from one to four clear threads beyond the nuts when tightened up. The shoulders of the heads shall be truly concentric with the axes of the bolts.

Coachscrews shall be of adequate length to penetrate the various members to be joined together, and shall have a large domed head with a square nut projecting from it. The heads shall be truly concentric with the axes of the coachscrews.

Dog spikes for fixing flat-bottom rails shall be of mild steel, 16 mm × 15 mm ($\frac{5}{8}$ in. × $\frac{9}{16}$ in.) in section and 125 mm (5 in.) long with cup-shaped heads and be suitably pointed for driving. The head shall be formed with a 20 mm ($\frac{3}{4}$ in.) projection with the required inclination to give a satisfactory grip on the rail, and with suitable ears to facilitate withdrawal.

After manufacture and before any rusting has taken place, bolts, nuts,

This clause covers all the components needed for fixing the rails. The dimensions of the various items will be obtained from the Drawings. Common sizes are for fishbolts 24 mm ($\frac{15}{16}$ in.) in diameter and 120 mm ($4\frac{3}{4}$ in.) long and for chairbolts 22 mm ($\frac{7}{8}$ in.) in diameter and 180 mm ($7\frac{1}{4}$ in.) long.

Reference to B.S. 64 enables the length of the specification clause to be reduced considerably. B.S.64 provides for 22 mm ($\frac{7}{8}$ in.) and 24 mm ($\frac{15}{16}$ in.) bolts and nuts for use with standard bull-head rails, and a range of ten bolts and nuts from 12 mm ($\frac{1}{2}$ in.) to 28 mm ($1\frac{1}{8}$ in.) for use with standard flat-bottom rails. This Standard also specifies the quality of steel and the tests on the finished bolt, as well as giving requirements for weight margins, marking, gauging and protection.

Note the application of hot linseed oil to facilitate removal and prevent rusting.

coachscrews, etc., shall be heated and dipped in hot linseed oil.

Cast iron separators

Cast iron separators, securing check rails to running rails, shall be of approved pattern and of the required size to fit accurately into the webs of the rails and of the appropriate width to give a clearance of 45 mm (1¾ in.) between the inner edges of the running and check rails. The separators shall be holed for 22 mm (⅞ in.) diameter distance bolts.

Cast iron separators are needed to hold the check rail in the correct position relative to the running rail on sharp curves.

Bearing plates

Mild steel bearing plates, 300 mm × 200 mm × 12 mm thick (12 in. × 8 in. × ½ in. thick), shall be placed between the rails and sleepers on each side of rail joints where directed by the Engineer. Bearing plates shall comply with B.S. 751: Steel Bearing Plates for Flat bottom Railway Rails, and each plate shall be suitably holed to permit four fixings being screwed or driven into the sleeper to hold the rail.

Bearing plates are used to give additional support to rails at joints. B.S. 751 specifies the quality of the steel, the tests to be performed, the method of manufacture, holing, freedom from defects, and branding, cleaning and dipping.

Chairs and keys

Chairs shall be of cast iron, weighing not less than 20 kg (46 lb) each, to take bull-head rails, and of a pattern approved by the Engineer. Chairs shall be of the three-hole type and they shall be secured to the sleepers with three coachscrews or spikes or with three bolts inserted through the full thickness of the sleeper, with the nut screwed down tight on to the chair.

Chairs are needed to fix each bull-head rail to each sleeper. Keys are used to wedge the rails in the chairs. Oak keys of good quality English oak, chamfered and tapered as required are an alternative to steel spring keys.

Special double cast iron

Keys shall be steel spring keys to railway standard pattern.

chairs will be needed with bull-head rails where a check rail is used.

Laying Permanent Way

Laying ballast

The ballast shall be laid after the formation has been brought to the correct level and profile and well consolidated with a roller weighing not less than 5 tonnes (5 tons), and be cleared of all rubbish and loose material. The width of the ballast for a single track shall be 3·5 m (11 ft), and the level of the formation after consolidation shall be 450 mm (18 in.) below the top of rail level.

Prior to the laying of sleepers, bottom ballast shall be laid to a consolidated thickness of 150 mm (6 in.). The permanent way shall then be laid and the sleepers packed up with top ballast for a width of 375 mm (15 in.) on each side of each rail. After the rails have been accurately adjusted, lined and surfaced, the top ballast shall be filled to the correct dimensions and neatly trimmed and boxed flush with the sleepers. Any settlement of the ballast prior to the expiration of the maintenance period shall be made good at the Contractor's expense and to the approval of the Engineer.

Ballast is laid in two layers, the bottom ballast of larger gauge material being consolidated to a depth of about 150 mm (6 in.) to receive the sleepers. The remaining space to the top of the sleepers is made up with top ballast of smaller material.

The accompanying specification clauses detail the ballast-laying process quite fully, and include making up any areas which have settled by the end of the maintenance period.

Laying track

The rails shall be accurately laid to line and level, to a gauge of $1\frac{1}{2}$ m (4 ft $8\frac{1}{2}$ in.) and to the true radii of the respective curves, with such super-elevation

The track must be laid to the lines, levels, curves and super-elevation shown on the Drawings or as directed by

on the outer rail on curves as required by the Engineer, and no addition to billed rates for tracklaying will be allowed to cover any extra cost involved in setting out and laying the track to the required elevation and curvature.

On straight sections of line the rail joints shall be positioned exactly opposite one another, while on curves the lengths of rails shall be so arranged that no joint shall have a lead of more than 100 mm (4 in.) over the joint of the opposite rail. The Contractor shall, where necessary to suit these conditions, cut the rails and bore new holes for the fishbolts and the cost of this work shall be included in the tracklaying rates. On curves less than 240 m (12 chains) in radius, the rails, prior to laying, shall be set by the Contractor to the required curvature with the use of a press.

The sleepers on each side of rail joints shall be placed close to the joints. The remainder of the sleepers in each length of rail shall be spaced equidistantly and they shall be so arranged that the distance between the centres of sleepers shall not exceed 750 mm (2 ft 6 in.).

The rails shall be laid as nearly as possible in correct alignment before they are secured to sleepers or chairs. One of the rails shall then be fishjointed and secured with the required number of fixings, keys, etc., as shown on the Drawings, after which the other rail shall be laid in its correct position with the use of a standard gauge, and then secured progressively to the sleepers or chairs. Metal slips, 7 mm ($\frac{5}{16}$ in.) thick, shall be inserted in the rail joints to provide expansion spaces and shall be kept in the joints until the rails have been lined

the Engineer. Furthermore, the track must be laid accurately and precisely. Billed rates for tracklaying must cover all the work entailed.

Sleepers are normally placed close to the joint on either side of rail joints, although an alternative procedure is to use a larger section sleeper in this position (often 300 mm × 125 mm (12 in. × 5 in.)). The spacing of sleepers (centre to centre) is usually about 750–850 mm (30–34 in.).

Note the sequence of operations adopted for tracklaying work as detailed in the accompanying specification clauses. There are three principal methods of fixing rails, using bolts and nuts, chairscrews, or spikes. With flatbottom rails no chairs will be needed as the rails will be fixed direct to the sleepers.

Bull head rails have a head and base or foot of the same width (about 70 mm ($2\frac{3}{4}$ in.)) and a height of about 140 mm ($5\frac{1}{2}$ in.). With flat bottom rails the width of the base is just over twice the width of the head. For example, a common section, weighing 47 kg/m (95 lb/lin. yd), has a base width of 140 mm ($5\frac{9}{16}$ in.), a head width of 65 mm ($2\frac{11}{16}$ in.) and a height of 145 mm ($5\frac{13}{16}$ in.). The bullhead type of rail is the most

and secured. Closing lengths of rail shall not be less than 4·5 m (15 ft) and all cuts in rails shall be square and clean. The Contract rates shall include for all cutting and waste resulting from the tracklaying. All the tracklaying shall be carried out in accordance with present-day first class railway practice.

Unless otherwise specified, there shall be at least six steel fixings securing the rails or chairs to each sleeper (i.e. at least three to each rail or chair). The fixings may consist of bolts and nuts, coachscrews or spikes, or a combination of them in the case of flat bottomed rails, all as shown on the Drawings or as directed by the Engineer. The bolts and coachscrews must be screwed tight and the spikes hammered home until a firm bearing on the rails or chairs is obtained, all to the approval of the Engineer.

Any loose coachscrews or spikes that have been rejected by the Engineer shall be removed and new holes drilled in the sleepers. New coachscrews or spikes shall then be screwed or driven at the Contractor's expense and to the approval of the Engineer. Steel spring keys shall be used to secure the rails to each chair and they shall be driven securely home after the rails have been gauged and lined.

Should any part of the permanent way settle, move or stretch prior to the expiration of the maintenance period, the Contractor shall perform the necessary remedial works to ensure that the track is left to the correct level, line and gauge, at his own expense and to the approval of the Engineer.

commonly used in this country.

Steel and precast concrete sleepers are particularly useful in tropical and sub-tropical regions, where timber sleepers are subject to attack by white ants.

Check rails

Check rails shall be provided on all curves of radius sharper than 120 m (6 chains) or where directed by the Engineer. They shall consist of a rail of the same section as the running rail, except that in the case of flat-bottom rails one flange shall be planed off as necessary to allow 45 mm (1¾ in.) clearance between the check rail and the running rail for the flanges of wheels.

Where bull-head rails are used, the check rails shall be secured to sleepers with special cast iron chairs, and in the case of flat-bottom rails by dog spikes or coachscrews and cast iron separators, and bolts and nuts between the webs of running and check rails at 1·5 m (5 ft) centres.

Check rails are needed on tracks laid to sharp curves as an additional safeguard to super-elevation to prevent locomotives leaving the track due to centrifugal force.

The form of check rail and the method of fixing varies according to whether the track is of bull head or flat bottom rails.

Points and crossings

Points and crossings, together with check railing, point rods and lever boxes complete, shall be manufactured by a firm of repute in this class of work. The rails shall be of the same section as previously specified. Prices shall include for all cutting and waste.

Points and crossings shall be accurately made to the particulars and drawings prepared or approved by the Engineer, and before construction is commenced the Contractor shall submit detailed working drawings to the Engineer for approval. Points and crossings shall be carefully and accurately laid to ensure the safe and smooth running of traffic.

Hand-lever boxes and tie rods of a pattern approved by the Engineer shall

With these more specialised forms of equipment, the Contractor is usually required to produce detailed drawings based on the information supplied by the Engineer. In fact the supply and fixing of points, crossings, buffer stops and associated items are frequently covered by a prime cost item in the Bill of Quantities. It is essential that this equipment should be in sound working order and produce safe running conditions.

be provided to work the points. They shall be properly fitted up and securely fixed to a pair of timber sleepers, all to the approval of the Engineer. Lever boxes and tie rods shall be painted with two coats of red lead paint and all moving parts and sliding plates shall be well greased and left in good working order.

Measurement of railway work

The billed rates for the supply of railway materials shall include all the costs involved in the supply, transporting, unloading, handling, stacking, storing and protection of the materials on the site. Payment will be made on the net theoretical or calculated weights of materials actually used in the Works in accordance with the Drawings or as directed by the Engineer.

The billed rates for ballasting shall include the cost of packing, surfacing, trimming and maintaining. No allowance will be made beyond the required dimensions for the extra material used in making up any settlement. The ballast displaced by sleepers will be deducted.

The billed rates for tracklaying shall include the cost of handling, selecting, fixing, lining and gauging rails, and also bending rails in the case of curved track, the handling and laying of transverse or block type sleepers, and all expenses connected with laying and fixing the permanent way complete, other than ballasting. The measurement of tracklaying will be taken in metres (linear yards) of completed track (two rails, all fixings and sleepers, etc.). The billed rates for check rails shall include the cost of handling, selecting and bending

It is advisable to indicate exactly what the billed rates for supply of railway equipment and ballasting and laying track, check rails and points and crossings are to cover, supplementing the information given in the Standard Method of Measurement of Civil Engineering Quantities. For instance, tracklaying is measured by the metre (yard) of track complete, including the assembly of all the component items. The billed items for points and crossings are also comprehensive items.

Payment for steel rails and other metal work, will be on the basis of theoretical or calculated weights as detailed by the manufacturer, and not on the basis of the actual weight of metal used on the job. The separation of the supply and the fixing of items is peculiar to railway construction work. Points are often referred to as switches.

rails, fixing with keys, spikes or coach-screws, drilling running and check rails for distance pieces, and fixing bolts and separators.

The billed rates for supply of points and crossings shall include for the supply of all necessary materials and for their manufacture complete, with the check rails, cast iron separators, bearing and joint plates, lever boxes, tie rods, etc., whilst the billed rates for laying points and crossings shall include the delivery, handling, laying in position, jointing up to the permanent way and fixing of all check rails, lever-boxes, etc., greasing and painting, and all cutting, drilling, machining and fitting of rails, rods and timbers which may be necessary for the complete erection of the points and crossings, measured as 'extra over' the laying of plain track.

Appendix I

LIST OF BRITISH STANDARD CODES OF PRACTICE RELATING
TO CIVIL ENGINEERING WORK

CP

3   Code of basic data for the design of buildings.
     Chapter   I   Lighting.
       ,,   II   Thermal insulation.
       ,,   III   Sound insulation and noise reduction.
       ,,   IV   Precautions against fire.
       ,,   V   Loading.
       ,,   VII   Engineering and utility services.
       ,,   VIII   Heating and thermal insulation.
       ,,   IX   Durability.
93   The use of safety nets on constructional works.
94   Demolition.
97   Metal scaffolding.
98   Preservative treatments for constructional timber.
99   Frost precautions for water services.
101   Foundations and substructures for non-industrial buildings of not more than four storeys.
102   Protection of buildings against water from the ground.
110   The structural use of concrete.
111   Structural recommendations for loadbearing walls.
112   The structural use of timber.
114   Structural use of reinforced concrete in buildings.
115   The structural use of prestressed concrete in buildings.
116   The structural use of precast concrete.
117   Composite construction in structural steel and concrete.
118   The structural use of aluminium.
121 Part 1   Brick and block masonry.
121.201   Masonry—walls ashlared with natural stone or with cast stone.
121.202   Masonry—rubble walls.
122   Walls and partitions of blocks and slabs.
123.101   Dense concrete walls.
142   Slating and tiling.
143   Sheet roof and wall coverings.
144 Part 3   Built-up bitumen felt.
144 Part 4   Mastic asphalt.
145   Glazing systems.

151	Doors and windows including frames and linings.
152	Glazing and fixing of glass for buildings.
153	Windows and rooflights.
199	Roof deckings.
231	Painting of buildings.
297	Precast concrete cladding (non-loadbearing).
298	Natural stone cladding (non-loadbearing).
301	Building drainage.
302	Small sewage treatment works.
304	Sanitary pipework above ground.
308	Drainage of roofs and paved areas.
310	Water supply.
413	Ducts for building services.
1004	Road lighting.
2001	Site investigations.
2003	Earthworks.
2004	Foundations.
2005	Sewerage.
2007	Design and construction of reinforced and prestressed concrete structures for the storage of water and other aqueous liquids.
2008	Protection of iron and steel structures from corrosion.
2010	Pipelines.
2011	Safety precautions in the construction of large diameter bore holes for piling and other purposes.

Appendix II

LIST OF BRITISH STANDARDS RELATING TO CIVIL ENGINEERING WORK

BS

4	Structural steel sections.
9	Bull head railway rails.
10	Flanges and bolting for pipes, valves and fittings.
11	Flat bottom railway rails.
12	Portland cement (ordinary and rapid-hardening).
18	Methods for tensile testing of metals.
47	Steel fishplates for bull head and flat bottom railway rails.
63	Single-sized roadstone and chippings.
64	Steel fishbolts and nuts for railway rails.
65 & 540	Clay drain and sewer pipes including surface water pipes and fittings.
76	Tars for road purposes.
78	Cast iron spigot and socket pipes (vertically cast) and spigot and socket fittings.
105	Light and heavy bridge-type railway rails.
144	Coal tar creosote for the preservation of timber.
146	Portland blastfurnace cement.
153	Steel girder bridges.
217	Red lead for paints and jointing compounds.
275	Dimensions of rivets 10 mm to 45 mm ($\frac{1}{2}$ in. to $1\frac{3}{4}$ in.) diameter.
308	Engineering drawing practice.
327	Power-driven derrick cranes.
340	Precast concrete kerbs, channels, edgings and quadrants.
357	Power-driven travelling jib cranes (rail-mounted low carriage type).
368	Precast concrete flags.
373	Testing small clear specimens of timber.
417	Galvanised mild steel cisterns and covers, tanks and cylinders.
434	Bitumen road emulsion (anionic and cationic).
435	Granite and whinstone kerbs, channels, quadrants and setts.
437	Cast iron spigot and socket drain pipes and fittings.
449	The use of structural steel in building.
459	Doors (various types).
486	Asbestos cement pressure pipes.
497	Cast manhole covers, road gully gratings and frames for drainage purposes.

BS

499	Welding terms and symbols.
500	Steel railway sleepers for flat bottom rails.
505	Road traffic signals.
534	Steel pipes fittings and specials for water, gas and sewage.
539	Dimensions of fittings for use with clay drain and sewer pipes.
544	Linseed oil putty for use in wooden frames.
556	Concrete cylindrical pipes and fittings including manholes, inspection chambers and street gullies.
565	Glossary of terms relating to timber and woodwork.
594	Rolled asphalt (hot process) for roads and other paved areas.
598	Sampling and examination of bituminous mixtures for roads and buildings.
599	Methods of testing pumps.
616	Methods for sampling coal tar and its products.
638	Arc welding plant, equipment and accessories.
639	Covered electrodes for the manual metal-arc welding of mild steel and medium-tensile steel.
648	Schedule of weights of building materials.
706	Sandstone kerbs, channels, quadrants and setts.
729	Hot dip galvanised coatings on iron and steel articles.
743	Materials for damp-proof courses.
747	Roofing felts.
750	Underground fire hydrants and dimensions of surface box openings.
751	Steel bearing plates for flat bottom railway rails.
778	Steel pipes and joints for hydraulic purposes.
785	Hot rolled bars and hard drawn steel wire for reinforcement of concrete.
802	Tarmacadam with crushed rock or slag aggregate.
812	Methods for sampling and testing of mineral aggregates, sands and fillers.
873	The construction of road traffic signs and internally illuminated bollards.
877	Foamed or expanded blastfurnace slag lightweight aggregate for concrete.
879	Water well casing.
881, 589	Nomenclature of commercial timbers, including sources of supply.
882, 1201	Aggregates from natural sources for concrete (including granolithic).
890	Building lime.
892	Glossary of highway engineering terms.
913	Wood preservation by means of pressure creosoting.
915	High alumina cement.
952	Classification of glass for glazing and terminology for work on glass.
988, 1076, 1097, 1451	Mastic asphalt for building (limestone aggregate).
1047	Air-cooled blast furnace slag coarse aggregate for concrete.
1130	Schedule of cast iron drain fittings, spigot and socket type, for use with drain pipes to B.S. 437.
1136	Mild steel refuse storage containers.
1139	Metal scaffolding.
1142	Fibre building boards.
1143	Salt-glazed ware pipes with chemically resistant properties.

232

BS

1151 Form of time and wages sheet and pay packet for the building and civil engineering contracting industries.

1161, 1410, 1418 Aluminium and aluminium alloy sections.

1162 Mastic asphalt for building (natural rock asphalt aggregate).

1165 Clinker aggregate for concrete.

1178 Milled lead sheet and strip for building purposes.

1180 Concrete bricks and fixing bricks.

1186 Quality of timber and workmanship in joinery.

1191 Gypsum building plasters.

1194 Concrete porous pipes for under-drainage.

1196 Clayware field drain pipes.

1198–1200 Building sands from natural sources.

1202 Nails.

1203 Synthetic resin adhesives (phenolic and aminoplastic) for plywood.

1208 Semi-rotary pumps, hand operated, double acting for water.

1210 Wood screws.

1211 Centrifugally cast (spun) iron pressure pipes for water, gas and sewage.

1217 Cast stone.

1218 Sluice valves for waterworks purposes.

1230 Gypsum plasterboard.

1239–40 Lintels (cast concrete and natural stone).

1241 Tarmacadam and tar carpets (gravel aggregate).

1242 Tarmacadam 'tarpaving' for footpaths, playgrounds and similar work.

1243 Metal ties for cavity wall construction.

1245 Metal door frames (steel).

1247 Manhole step irons (malleable cast iron).

1282 Classification of wood preservatives and their methods of application.

1305 Batch type concrete mixers.

1308 Concrete street lighting columns.

1336 Knotting.

1377 Methods of testing soils for civil engineering purposes.

1387 Steel tubes and tubulars suitable for screwing to B.S. 21 pipe threads.

1426 & 3461 Surface boxes for gas and waterworks purposes.

1438 Media for biological percolating filters.

1446 Mastic asphalt (natural rock asphalt fine aggregate) for roads and footways.

1447 Mastic asphalt (limestone fine aggregate) for roads and footways.

1455 Plywood manufactured from tropical hardwoods.

1521 Waterproof building papers.

1553 Graphical symbols for general engineering.

1563 Cast iron sectional tanks (rectangular).

1564 Pressed steel sectional tanks (rectangular).

1579 Connectors for timber.

1621 Bitumen macadam with crushed rock or slag aggregate.

1622 Winter gritters for roads.

1623 Hand-rollers for road and constructional engineering.

1634 Dimensions of stoneware pipes and pipe fittings for chemical purposes.

1639 Methods for bend testing of metals.

1676 Heaters for tar and bitumen (mobile and transportable).

1690 Cold asphalt.

BS	
1703	Refuse chutes.
1707	Hot binder distributors for road surface dressing.
1710	Identification of pipelines.
1722	Fences.
1788	Street lighting lanterns for use with electric lamps.
1849	Steel columns for street lighting.
1853	Tubular fluorescent lamps for general lighting service.
1856	General requirements for the metal-arc welding of mild steel.
1860	Structural timber – measurement of characteristics affecting strength.
1881	Methods of testing concrete.
1924	Methods of test for stabilized soils.
1926	Ready-mixed concrete.
1984	Gravel aggregates for surface treatment (including surface dressings) on roads.
2015	Glossary of paint terms.
2028, 1364	Precast concrete blocks.
2035	Cast iron flanged pipes and flanged fittings.
2040	Bitumen macadam with gravel aggregate.
2494	Rubber joint rings for gas mains, water mains and drainage purposes.
2499	Hot applied joint sealants for concrete pavements.
2521 & 2523	Lead based priming paints.
2525–27	Undercoating and finishing paints for protective purposes (white lead-based).
2539	Preferred dimensions of reinforced concrete structural members.
2569	Sprayed metal coatings.
2573	Permissible stresses in cranes.
2591	Glossary for valves and valve parts (for fluids).
2594	Horizontal mild steel welded storage tanks.
2596	Components of crawler tractors and earth moving equipment.
2691	Steel wire for prestressed concrete.
2760	Pitch-impregnated fibre pipes and fittings for below and above ground drainage.
2787	Glossary of terms for concrete and reinforced concrete.
2853	The design and testing of steel overhead runway beams.
2994	Cold rolled steel sections.
3049	Pedestrian guard rails (metal).
3051	Coal tar creosotes for wood preservation (other than creosotes to B.S. 144).
3083	Hot-dipped galvanised corrugated steel sheets for general purposes.
3100	Steel castings for general engineering purposes.
3136	Cold emulsion spraying machines for roads.
3138	Glossary of terms used in work study.
3139	High strength friction grip bolts for structural engineering.
3148	Tests for water for making concrete.
3178	Playground equipment for parks.
3224	Lighting fittings for civil land aerodromes.
3247	Salt for spreading on highways for winter maintenance.
3251	Hydrant indicator plates.
3262	Road marking materials.
3294	The use of high strength friction grip bolts in structural steelwork.

BS	
3327	Stationery for quantity surveying.
3373	Wrought magnesium alloys for general engineering purposes – bars, section and tubes including extruded forging stock.
3429	Sizes of drawing sheets.
3436	Ingot zinc.
3505	Unplasticized PVC pipe for cold water services.
3572	Access fittings for chimneys and other high structures in concrete or brickwork.
3656	Asbestos-cement pipes, joints and fittings for sewerage and drainage.
3680	Methods of measurement of liquid flow in open channels.
3681	Methods for the sampling and testing of lightweight aggregates for concrete.
3690	Bitumens for road purposes.
3698	Calcium plumbate priming paints.
3699	Calcium plumbate for paints.
3717	Asbestos-cement decking.
3767	Low pressure sodium vapour lamps.
3797	Lightweight aggregates for concrete.
3798	Coping units (of clayware, unreinforced cast concrete, unreinforced cast stone, natural stone and slate).
3809	Wood wool permanent formwork and infill units for reinforced concrete floor and roofs.
3882	Recommendations and classification for top soil.
3892	Pulverised-fuel ash for use in concrete.
3921	Clay bricks and blocks.
3969	Recommendations for turf for general landscape purposes.
3981	Iron oxide pigments for paints.
3989	Aluminium street lighting columns.
3998	Recommendations for tree work.
4011	Recommendations for co-ordination of dimensions in building. Basic sizes for building components and assemblies.
4016	Building papers (breather type).
4027	Sulphate-resisting Portland cement.
4043	Recommendations for transplanting semi-mature trees.
4047	Grading rules for sawn home grown hardwood.
4072	Wood preservation by means of water-borne copper/chrome/arsenic compositions.
4076	Steel chimneys.
4082	External dimensions for vertical in-line centrifugal pumps.
4101	Concrete unreinforced tubes and fittings with ogee joints for surface water drainage.
4108	Pitch-impregnated fibre conduit.
4118	Glossary of sanitation terms.
4229	Recommendations for metric sizes of non-ferrous and ferrous bars.
4232	Surface finish of blast-cleaned steel for painting.
4248	Supersulphated cement.
4251	Truck type concrete mixers.
4261	Glossary of terms relating to timber preservation.
4300	Specification (supplementary series) for wrought aluminium and aluminium alloys for general engineering purposes.

BS	
4346	Joints and fittings for use with unplasticized PVC pressure pipes.
4360	Weldable structural steels.
4374	Sills of clayware, cast concrete, cast stone, slate and natural stone.
4408	Recommendations for non-destructive methods of test for concrete.
4428	Recommendations for general landscape operations (excluding hard surfaces).
4449	Hot rolled steel bars for the reinforcement of concrete.
4461	Cold worked steel bars for the reinforcement of concrete.
4466	Bending dimensions and scheduling of bars for the reinforcement of concrete.
4469	Car parking control equipment.
4471	Dimensions for softwood.
4482	Hard drawn mild steel wire for the reinforcement of concrete.
4483	Steel fabric for the reinforcement of concrete.
4484	Measuring instruments for constructional works.
4485	Water cooling towers.
4486	Cold worked high tensile alloy steel bars for prestressed concrete.
4504	Flanges and bolting for pipes, valves and fittings.
4508	Thermally insulated underground piping systems.
4519	External dimensions for horizontal end-section centrifugal pumps.
4521	Railway turnouts for private users.
4533	Electric luminaires (lighting fittings).
4543	Factory-made insulated metal chimneys.
4550	Methods of testing cement.
4551	Methods of testing mortars and specification for mortar testing sand.
4604	The use of high strength friction grip bolts in structural steelwork.
4619	Heavy aggregates for concrete and gypsum plaster.
4620	Rivets for general engineering purposes.
4625	Prestressed concrete pressure pipes (including fittings).
4627	Glossary of terms relating to types of cements, their properties and components.
4660	Unplasticized PVC underground drain pipes and fittings.
4721	Ready-mixed lime: sand for mortar.
4756	Ready-mixed aluminium priming paints for woodwork.
4848	Hot-rolled structural steel sections.
4868	Profiled aluminium sheet for building.
4873	Aluminium alloy windows.
4887	Mortar plasticizers.
4943	Co-ordinating sizes for corrugated sheet materials used in building.
4949	Glossary of terms relating to building performance.
4962	Performance requirements for plastics pipe for use as light sub-soil drains.
4978	Timber grades for structural use.
4987	Coated macadam for roads and other paved areas.
5041	Fire hydrant systems equipment.
5056	Copper naphthenate wood preservatives.
5080	Methods for test for structural fixings in concrete and masonry.
5082	Water-thinned priming paints for wood.

Appendix III

TYPICAL PROGRAMME OF WORKS COVERING TUNNELS AND SHAFTS TO A CIRCULATING WATER SYSTEM TO A POWER STATION

Riverworks & outfall shaft constn.

Penstock shaft constn. (excl. lining).

Main station shaft constn. (excl. lining)

Tunnel lining M.S. to P. shafts

Tunnel M.S. to P. Continue tunnel from P. shaft towards O. shaft

Shafts shaft constn. (excl. lining).

Lining to shafts & tunnels

Tunnel from O. shaft towards P. shaft

Intake shaft

Pumphouse shaft (excl. lining)

Intake tunnel const.

Penstocks, etc.

Prelim. site work

3-month test period

31st. Proposed date of commencement of contract

1st. Commence site preliminaries

30th. Complete site preliminaries

1st. Commence main station shaft, penstock shaft & outfall shaft

30th. Complete main station shaft (excluding lining)

11th. Commence tunnel from main station shaft to penstock shaft (excluding lining)

31st. Complete penstock shaft (excluding lining)

31st. Complete tunnel from main station shaft to penstock shaft

1st. Commence driving from penstock shaft

1st. Access for aeration and associated plant

31st. Complete outfall shaft (excluding lining)

11th. Access to intake shaft and to pumphouse shaft
Commence driving from outfall shaft towards penstock shaft

1st. Commence intake and pumphouse shafts

31st. Complete main outfall tunnel (excluding lining)

10th. Complete intake and pumphouse shafts (excluding lining)

21st. Commence intake tunnel

30th. Complete lining to shafts and outfall tunnel

31st. Complete intake tunnel (excluding lining)

4th. (excl. lining) transfer shield

31st. Complete outfall tunnel, including lining, shafts and penstocks

28th. Complete lining to shafts and intake tunnel

31st. Complete intake tunnel, shafts and penstocks

30th. Commissioning of Set

MONTH	APL.	MAY	JUNE	JLY.	AUG.	SEPT.	OCT.	NOV.	DEC.	JAN.	FEB.	MAR.	APL.	MAY	JUNE	JLY.	AUG.	SEPT.	OCT.	NOV.	DEC.	JAN.	FEB.	MAR.	APL.	MAY	JUNE
PERIOD	II		III			IV				I			II		III			IV				I			II		
YEAR			1967												1968									1969			

PROPOSED PROGRAMME OF WORKS

POWER STATION ~ CIRCULATING WATER MAIN TUNNELS AND SHAFTS

APPENDIX III

Appendix IV

METRIC CONVERSION TABLE

Length
1 in. $=25 \cdot 44$ mm (approximately 25 mm)

then $\dfrac{\text{mm}}{100} \times 4 = $ inches

1 ft $=304 \cdot 8$ mm (approximately 300 mm)
1 yd $=0 \cdot 914$ m (approximately 910 mm)
1 mile $=1 \cdot 609$ km (approximately $1^3/_5$ km)
1 m $=3 \cdot 281$ ft $=1 \cdot 094$ yd (approximately
$\qquad 1 \cdot 1$ yd) (10 m $=11$ yd approximately)
1 km $=0 \cdot 621$ mile ($^5/_8$ mile approximately)

Area
1 ft^2 $=0 \cdot 093$ m^2
1 yd^2 $=0 \cdot 836$ m^2
1 acre $=0 \cdot 405$ ha (1 ha or hectare $=10\,000$m^2)
1 mile^2 $=2 \cdot 590$ km^2
1 m^2 $=10 \cdot 746$ ft^2 $=1 \cdot 916$ yd^2 (approximately
$\qquad 1 \cdot 2$ yd^2)
1 ha $=2 \cdot 471$ acres (approximately $2^1/_2$ acres)
1 km^2 $=0 \cdot 386$ mile^2

Volume
1 ft^3 $=0 \cdot 028$ m^3
1 yd^3 $=0 \cdot 765$ m^3
1 m^3 $=35 \cdot 315$ ft^3 $=1 \cdot 308$ yd^3 (approximately
$\qquad 1 \cdot 3$ yd^3)
1 ft^3 $=28 \cdot 32$ litres (1000 litres $=1$ m^3)
1 gal $=4 \cdot 546$ litres
1 litre $=0 \cdot 220$ gal (approximately $4^1/_2$ litres to
$\qquad$ the gallon)

Pressure
1 lbf/in.^{2} $=0 \cdot 007$ N/mm^2
1 lbf/ft^2 $=47 \cdot 88$ N/m^2
1 tonf/ft^2 $=107 \cdot 3$kN/m^2

Mass
1 lb $=0 \cdot 454$ kg (kilogram)
1 cwt $=50 \cdot 80$ kg (approximately
$\qquad$ 50 kg)
1 ton $=1 \cdot 016$ tonne (1 tonne $=$
$\qquad 1000$ kg $=0 \cdot 984$ ton)
1 kg $=2 \cdot 205$ lb (approximately
$\qquad 2^1/_5$ lb)

Density
1 lb/ft^3 $=16 \cdot 019$ kg/m^3
1 kg/m^3 $=0 \cdot 062$ lb/ft^3

Velocity
1 ft/s $=0 \cdot 305$ m/s
1 mile/h $=1 \cdot 609$ km/h

Energy
1 therm $=105 \cdot 506$ MJ (megajoules)
1 Btu $=1 \cdot 055$ kJ (kilojoules)

Thermal conductivity
1 Btu/ft^{2}h$^\circ$F $=5 \cdot 678$ W/m^2°C
$\qquad$ (where W$=$watt)

Temperature
$x\,^\circ$F $=^5/_9(x-32)\,^\circ$C
$x\,^\circ$C $=^9/_5x \quad 32\,^\circ$F
$0\,^\circ$C $=32\,^\circ$F (freezing)
$5\,^\circ$C $=41^\circ$F
$10\,^\circ$C $=50\,^\circ$F (rather cold)
$15\,^\circ$C $=59^\circ$F
$20\,^\circ$C $=68\,^\circ$F (quite warm)
$25\,^\circ$C $=77^\circ$F
$30\,^\circ$C $\quad 86\,^\circ$F (very hot)

Index

ACCESS TO SITE 36
Accommodation for employees 45
Acts and regulations 32
Advertising 37–8
Aggregate
  coarse 68
  fine 67–8
  for granular bases 144
  for surface dressing 144
  samples 69
  stocks 69
Air chambers 213
Aluminium sheeting 124–5
Angle fillet 100
Asbestos cement
  pipes 174
  sheeting 124
Ashes 143
Asphalt 100–1
  cold 145–6, 157
  footpaths 167–8
  rolled 145, 156–7

BACKFILL 54–5, 177–8
Ballast 217
  laying 222
Base
  lean concrete 152–3
  sub- 152
  water-bound granular 152
Benchings 189
Bending reinforcement 81
Bill of quantities 3–5
Bitumen
  cut-back 144

felt d.p.c. 93–4
macadam 145, 155–6
road emulsion 144
sheeting 101–2
Blinding coat 76
Bollards 133
Bolts 115, 117–18, 131–2
  to railway trackwork 220–1
  to segments 192
Borings 50
Boxsteps 189–90
Braces 129–30
Bricklaying 91–2
  in frosty weather 93
Bricks
  commons 89
  engineering 90, 174
  facing 90
  generally 89
Brickwork
  bonding 92
  faced 94
  labours to 92–3
  pointing 94
  reinforced 94–5
  to manholes 187–8
British Standards 21, 22–4

CABLE DUCTS 166
Capstans 134
Carpentry work 128
Cast iron specials 173–4, 201–2
Caulking joints to cast iron segments 192–3
Cement 67

Central Electricity Generating Board 3
Certificates 43–4
Chairs 221–2
Channel
    manhole 189
    road 148, 165
Check rails 225
Chimney shaft linings 95
Chlorination of mains 211
Coachscrews 220
Codes of Practice 22, 24–5
Cofferdams 62–3
Compacting
    concrete 161–2
    factor 72–3
Compressed air plant 59–60
Concrete
    blinding 76
    compacting 161–2
    consistency 72–3
    construction joints 76–7
    curing 79, 164
    expansion joints 77, 162–4
    flags 148, 168
    footpaths 168
    gauging 70–1
    in cold weather 78–9
    lean base 152–3
    lining to shaft and tunnel rings 194–5
    manholes 188–9
    mixes 69–70, 159–60
    mixing 71–2
    percolation tests 73–4
    piles 105–9
    pipes 173
    placing 74–6, 160–1
    precast 85–6
    prestressed 86–7
    protection to pipes 183–4
    records 80–1
    reinforcement 81–2
    roads 158–64
    shuttering 82–5
    sleepers 219
    slump 72–3
    surface finish 77–8
    tensioning 86–7
    test cubes 73
    transporting 74
    vibrated 76
Construction joints 76–7

Contract
    all-in 14
    bill of quantities 12
    cancellation of 30–1
    cost plus fixed fee 13
    cost plus fluctuating fee 13
    cost plus percentage 13
    documents 1
    drawings 5–6, 21, 30
    extent of 29–30
    form of 1–2
    General Conditions 2–3
    I.C.E. 7–9
    I.Struct.E. 9–12
    lump sum 12
    schedule 12–13
    target 13–14
Contractor to visit site 35–6
Co-ordination with other contractors 45
Coping stones 98
Costs, matters affecting 33–6
Cover
    to manhole 175, 191
    to reinforcement 82
Coverings to walls and roofs 124–5
Cramps 98–9
Creosoting
    piles 110
    timberwork 131
Crossings (railway) 225–6, 227
Cupboards 140
Curing
    concrete 79, 164
    piles 106–7

DAMAGE TO ADJOINING PROPERTIES 42
Damp-proof courses 93–4, 100–1
Demolition work 52
De-watering 57
Disposal of surplus soil 54
Diversion of services 38
Dog spikes 220
Doors 138–9
Dowels 98–9
    bars 146, 163
Drawings 5–6, 21, 30
Dredging 63–4
Dressed stonework 95–6
Dry rubble walling 98

EDGINGS 148, 165

Electric cable ducts 166
Electricity supply 34–5
Electrodes 119
Embankments 55–6
Employer's requirements 21
Engineering bricks 90, 174
Environment, Department of 3
Excavation 52–3
  backfilling 54–5, 177–8, 209
  excess 177
  for railways 216
  for roads 150
  of pipe trenches 54, 205
  of trenches and for manholes 175–6
  to be kept free of water 57, 177
Existing services 39, 208
Expansion joints 77, 162–4, 168

FABRIC REINFORCEMENT 146
Fencing 169–70
Fender piles 128–9
  rubber buffers to 132–3
Fertiliser 149, 169
Fibreglass scumboards 198
Fill 55–6
Filling to roads 151–2
Fishplates 219–20
Flags 148, 168
Flush doors 139
Footbridge 135–6
Footpath
  asphalt 167–8
  concrete 168
  edgings 148, 165
  flagged 148, 168
  surfacings 144–6
  tarmacadam 166–7
Form of agreement 2
Form of contract 1–2
Form of tender 6–7
Forms 147, 159
Formation preparation
  railways 216
  roads 150–1
Framing of joinery 137–8

GALVANISING 123–4
General Conditions of Contract 2, 29
General contractual matters 28–31
General requirements 28
General working requirements 44–7
Granular bases 144

Grass seed 149
  spreading 56
  verges 169
Guardrails
  timber 130
  tubular steel 121–2
Gully
  gratings 149
  pots 149
  road 165–6

HANDRAILS 121–2
Hardcore 143
Hedge removal 51
Hydrants 203–4
  chambers 214

I.C.E. CONDITIONS 7–9
Ironmongery 139
I.Struct.E. Conditions 9–12

JETTIES 127–35
  bollards 133
  capstans 134
  decking 130
  lighting installation 132
  mooring rings 134–5
  rescue chains 134
Joinery work 136–40
  quality of 137–8
Joint
  filler 146
  sealer 147
Jointing
  asbestos cement pipes 182
  cast iron segments 192–3
  clayware and concrete pipes 181
  flanged pipes 206
  iron pipes 181–2
  pipes generally 180–1
  pitch-fibre pipes 183
  precast concrete segments 193–4
  screw gland pipes 206

KERBS 147–8, 165
Keys 221–2
Knotting 140

LABOUR EXPENSES 33
Ladders 121, 190
Legal provisions 32–3

MANHOLE
  benchings and channels 189
  boxsteps 189–90
  brick 187–8
  covers 175, 191
  excavation 175–6
  ladders 190
  precast concrete 188–9
  safety bars 191
  safety chains 190
  step irons 175, 189
Marker plates 214
Masonry 95–100
Mastic asphalt 100–1
Materials descriptions 18–20, 43
Measurement 46
  railway work, 226–7
  steelwork 119–20
  timberwork 132
Media 196
  placing 197–8
  testing 197
Metric conversions 237
Mooring rings 134–5
Mortar 90–1

NAME PLATES 214
Navigation lights 132

OFFICE FOR RESIDENT ENGINEER 40–1
Open steel flooring 122

PAINTING
  steelwork 122–3, 140–1
  woodwork 140–1
Percolating filters
  distributors 195
  media 195–8
Percolation tests 73–4
Performance specification 55
Photographs 46–7
Pile
  casting 106
  concrete 105–9
  creosoting 110
  curing, stripping and stacking 106–7
  cutting off heads 111
  cutting steel sheet 112–13
  damaged or misplaced 113
  drilling 113
  fender 128–9

frames 108
handling 107
lengthening of 109
pitching and driving 108–9, 111,
  112–13
ready-made 107
reinforcement 105–6
rings 110–11
shoes 106, 110–11
steel sheet 112–13
tarring 110
timber 109–11
trial 107
Pipe trench
  backfilling 54–5, 177–8, 209
  excavation 54, 205
  reinstatement 209–10
Pipes
  asbestos cement 174
  building in 180
  chlorination of 211
  concrete 173
  concrete protection to 183–4
  cutting 183
  glazed vitrified clay 173
  in river crossings 208
  jointing 180–3, 206–7
  junction 183
  laying 179–80, 206–8
  loading and unloading 178–9, 205–6
  location 178
  pitch-fibre 174
  porous 174
  protective coatings on 202
  P.V.C. 202
  spun iron 173, 201
  steel 202
  support to 177
  testing 184–7, 210–11
  trench excavation 54, 205
Plant 42
  restricted use of 56
Plywood 139
Pointing of brickwork 94
Points (railway) 225–6, 227
Port regulations 32–3
Portland cement 67
Precast concrete 85–6
  manholes 188–9
Prestressed concrete 86–7
Price variations 30

Primers 140–1
Programme and progress record 47–8, 236
Protection of work 47
Protective coatings on pipes 202
Pumping 57

QUADRANTS 148, 165

RAILS 219
  check 225
  fixings 220–1
  laying 222–4
Railway
  ballast 217, 222
  bearing plates 221
  chairs 221–2
  check rails 225
  excavation 216
  fishplates 219–20
  formation preparation 216–17
  keys 221–2
  measurement 226–7
  points and crossings 225–6, 227
  rails 219
  separators 221
  sleepers 217–19
  tracklaying 222–4
Random rubble 97–8
Records 47–8
  concreting 80–1
Reinforcement 68–9, 81–2
  bending 81
  cover to 82
  fabric 146
  placing 81–2, 160–1
  to piles 105–6
Reinstatement of trench surfaces
  permanent 209–10
  temporary 209
Rescue chains 134
River authority regulations 32–3
River crossings 208
Riveting 118
Road
  asphalt 156–7
  base 143–4
  bitumen macadam 155–6
  channels 148, 165
  concrete 158–64

emulsion 144
filling 151–2
forming preparation 150–1
forms 147
kerbs 147–8, 165
lean concrete base 152–3
quadrants 148–165
sub-base 152
surface dressing 157–8
surfacing 144–6
tar 144
tarmacadam 154
water-bound granular base 152
Rubble walls 97–8

SAFETY
  bars 191
  chains 190
  precautions 33–4
Samples 43–4
  of aggregate 69
Sanitary conveniences 41
Screens 198–9
Scumboards 136
Separators 221
Sequence of works 34
Setting out 40
Sewage works
  filter distributors 195
  filter media 196–8
  screens 198–9
  scumboards 136, 198
Shaft
  excavation 60–1
  grouting 193
  jointing 192–4
  linings 191–5
  segments 191
Sheet piling 57–8
Shelving 139
Shield-driving 61–2
Shuttering 82–5
  design and construction of 82–3
  preparation of 84
  striking 84–5
  to beams and slabs 84
  to vibrated concrete 83
Side slopes, trimming, 56
Site
  clearance 51–2, 215–16
  investigations 21, 50–1

levels 51
tidiness 45–6
Skirtings
asphalt 101
wood 140
Sleepers
concrete 219
steel 219
timber 217–18
Slope formation 217
Sluice valves 203
chambers 211–13
Slump tests 72–3
Soil-cement 153
Specification
drafting clauses 18–20
general arrangement 17
general clauses 26–48
functions 16–17
sources of information 20–2
Squared rubble 97
Steel
bolting 117–18
corrugated sheeting 125
erection 117
fabrication 115–16
flooring 122
galvanising 123–4
guardrails 121–2
inspection and marking 116
ladders 121
measurement 119–20
painting 122–3
pipes 202
reinforcement, 68–9, 81–2, 105–6
riveting 118
sheet piles 112–13
sleepers 219
structural 115–20
testing 120
welding 118–19
Step irons 175–89
Steps 135
Stonework
cast 99–100
dressed 95–6
Stopping 140
Storage 37
Sub-base 152
Subsoil investigations 36
Supports for existing pipes 177

Surface
boxes 204
dressing 151, 157–8
finish of concrete 77–8
soil stripping 52, 150
water drainage 216
Suspension of works during bad weather
44

TANKING 100–1
Tar for surface dressing 144
Tarmacadam 144–5
footpaths 166–7
roads 154
Tarring
piles 110
timberwork 130–1
Telephone 41
Temporary works 38–42
Tender, sufficiency of 31
Tests 43–4
compaction 72–3
cubes 73
filter media 197
of steelwork 120
percolation 73–4
pipes 184–7, 210–11
slump 72–3
watertightness of tanks 80
Thrust blocks, 184
Ties to hollow walls, 92
Thrust blocks, 184
Timber 127–8
creosoting 131
cupboards 140
decking 130
doors 138–9
footbridge 135–6
guardrail 130
labours 128, 137–8
measurement 132
painting 140–1
piles 109–11
quality for joinery 136
scumboards 136
shelving 139
skirtings 140
sleepers 217–18
steps 135
tarring 130–1
windows 138

Timbering 57–8, 176–7
Tracklaying 222–4
Trade catalogues 22
Traffic control 38
Tree removal 51–2
Trial
  holes 51, 176
  piles 107
Tunnel
  grouting 193
  jointing 192–4
  linings 191–5
  segments 191–2
  work 58–62
Turf 52

USE
  of public highways 42
  of site 37

VALVE
  air 203
  hydrant 203–4
  keys 204
  marker plates 214

  sluice 203
  surface boxes 204
Valve chambers
  air 213
  hydrant 214
  sluice 211–13
  washout 213
Ventilating columns 195
Verges 169
Voussoirs 98

WALINGS 129–30
Washout valve chambers 213
Water 68
  content of concrete 72
  levels 51
  supply 35
Water-bound granular base 152
Waterproof underlay 146, 158
Welding 118–19
Wharves 127–35
Windows 138
Work prepared off site 44
Working area 37
Working rule agreement 33
Workmanship clauses 20, 43